Imitation of Life

Imitation of Life

How Biology Is Inspiring Computing

Nancy Forbes

The MIT Press
Cambridge, Massachusetts
London, England

This book was set in Sabon by SNP Best-set Typesetter Ltd., Hong Kong and was printed and bound in the United States of America.

Library of Congress Cataloging-in-Publication Data

Forbes, Nancy.
Imitation of life: how biology is inspiring computing / Nancy Forbes.
   p. cm.
Includes bibliographical references (p.).
ISBN 0-262-06241-0 (hc. : alk paper)
1. Biology.   2. Computer science.   I. Title.

QH307.2.F64   2004
570—dc22

                                                    2003064195

10 9 8 7 6 5 4 3 2 1

To my father,
Charles C. Forbes,
who nurtured my curiosity and love of science

and to Harvey Brooks,
who mentored me

Contents

Preface   ix

1   Artificial Neural Networks   1

2   Evolutionary Algorithms   13

3   Cellular Automata   25

4   Artificial Life   37

5   DNA Computation   51

6   Biomolecular Self-Assembly   67

7   Amorphous Computing   83

8   Computer Immune Systems   97

9   Biologically Inspired Hardware   113

10   Biology Through the Lens of Computer Science   139

Epilogue   155

Notes   159

Index   163

Preface

This book was born out of my fascination and involvement with science in the area of overlap between biology and computer science—two disciplines that on the surface may seem to have little commonality. My introduction to this field came several years ago, while working on a defense-related government research project whose objective was to explore unconventional ways of computing without the use of silicon. The scientists and engineers associated with this project constituted about as renowned and creative a cadre of multidisciplinary minds as could be found anywhere in the world, and they had been given relative freedom to explore and implement some truly original, even radical ideas for non-silicon-based information processing systems.

As a consequence, I learned about such revolutionary forms of information processing as quantum computing (based on the manipulation of atomic particles, such as electrons or nuclei, for carrying out the binary logic of computation), computing with strands of DNA, or how to construct digital logic gates inside living cells. As part of the thought process of reinventing computing, some researchers saw the need to return to the conceptual foundations of the discipline, reexamining computing's early models or such basic notions as the thermodynamical relationship between energy and information processing. Others explored how best to exploit the various advantages that information processing in different media would present over conventional silicon computers, such as an increase in parallel processing ability (DNA computing) or in speed (quantum computation); or whether there were aspects of biological systems, such as a swarm of bees or the mammalian brain, that could be co-opted as a basis for new models for algorithm development, or more efficient hardware.

In reexamining the theoretical underpinnings of computer science, it would have been impossible for these researchers to avoid bumping up against the ideas of the two most important early figures in the discipline, mathematicians John von Neumann and Alan Turing, whose work came to the fore roughly around the middle of the last century. So much has been written about them and the magnitude of their contributions to mathematics, computer science, and other areas, that to summarize it here in a few short sentences would be a daunting, if nearly impossible task. For the purposes of this preface, however, it may suffice to highlight one key fact: both men had more than a passing interest in the life sciences. At one point in his career, Turing's attention was drawn to the area of interface between mathematics, chemistry, and biology, and he wrote a definitive paper on chemical morphogenesis, the theory of growth and form in biology. When investigating how animal pigmentation develops such that some animals have spots instead of stripes, Turing postulated a chemical reaction-diffusion model that described the way in which different pigmentation cells interact and spread over the skin, darkening one area while leaving the other colorless. His equations remarkably revealed that the different patterns of pigmentation depended only on the size and shape of the surface area where they formed.

Following his work with high-speed calculating machines in the mid 1940s, Hungarian John von Neumann—mathematician, scientist, logician, and architect of the first digital computer—became very interested in the way biological systems process information. He began to read widely in neurophysiology, establish contacts in the biomedical community, and participate in interdisciplinary conferences on the topic.

One of these was the Hixon Symposium on the brain, which took place in September 1948, at the California Institute of Technology. Most of the symposium speakers were from the cognitive and physiological sciences, including such well-known researchers as Dr. Karl Lashley from Harvard (generally considered the father of neuropsychology), and Dr. Warren McCullough from the University of Illinois, known, with Walter Pitts, for first modeling the activity of neurons in the brain in terms of mathematical logic, thus laying the foundation for the field of artificial neural nets. Although some may have thought it incongruous to see a mathematician listed among the speakers, someone had the foresight to invite John von Neumann, probably because of his interest in biology and his renown as a logician.

At the Hixon Symposium, von Neumann gave what proved to be a seminal lecture, called "The General and Logical Theory of Automata." *Webster's Dictionary* defines automata as "systems or control mechanisms designed to follow automatically a predetermined sequence of operations or respond to encoded instructions." Although not explicitly defined in his talk, it is assumed von Neumann took automata to mean any system that uses or processes information as part of its self-regulating mechanisms. As examples, he pointed to the human nervous system and the computing machine, which could be compared as equivalent entities once both were abstracted and reduced down to their most essential parts and processes. He strongly believed that a general theory of automata that could apply both to machines and living systems had to be based on mathematical logic, although he recognized that, in reality, biological organisms were immeasurably more complex than computers, and additional study would be warranted to better understand the role of probability and self-replication in automata. (Parenthetically, given von Neumann's genius, it's not entirely surprising that his thoughts on the logic of self-replication in automata—the way in which a machine, if capable of reproducing itself, is organized to contain a complete description of itself, and then uses that information to create new copies of itself, while also passing on a copy of the description to its offspring—largely prefigured the discovery of the function and structure of DNA by Watson and Crick a few years later.)

In his Hixon lecture, von Neumann suggested that there is much the fledgling field of computing could learn from the study of nature.

Natural organisms are, as a rule, much more complicated and subtle, and therefore much less well understood in detail, than are artificial automata. Nevertheless, some regularities which we observe in the organization of the former may be quite instructive in our thinking and planning of the latter.[1]

For example, both living and mechanical automata process information via an "on/off" or digital switching mechanism. Vacuum tubes, the building blocks of early computers, turned on and off the flow of a current from electrodes, whereas neurons in the brain (whose mode of functioning had been described in terms of logical rules by McCullough and Pitts only a few years earlier) make use of an electrical potential difference across the cell membrane that stimulates or inhibits the firing of electrical impulses with switching at the synapse via neurotransmitter release. In both cases, these digital switching mechanisms form the basic elements (binary logic and neurophysiological activity) that enable the system as a whole to process information. Von Neumann felt that computers (composed of vacuum tubes as they were at the time) could probably never obtain the degree of complexity of natural automata because, in terms of size, spacing, organization, and the electrical properties of their switching elements, they used materials that were inferior to natural ones. Moreover, they had many fewer working parts, and were much more vulnerable to errors and malfunctions than biological systems that were redundant, self-diagnosing, and self-healing.

Perhaps more important, von Neumann had realized even earlier that nature had created the most powerful information processing system conceivable, the human brain, and that to emulate it would be the key to creating equally powerful man-made computers. As such, it served as his inspiration for the design of one of the earliest digital computers. The switching devices or logic gates for his computer were modeled on Pitts-McCullough neurons; he added other elements such as input/output devices, memory, data storage, and central processing units that also mirrored the brain's organizational structure for processing information. In fact, the vast majority of today's computers are based on this design and are called "von Neumann machines.[2]

Perhaps it's not completely fortuitous that several decades after von Neumann's Hixon Lecture, a third computing visionary, Seymour Cray, the father of the supercomputer, also expressed a belief that biology would have a significant influence on the future of information processing. In a speech in 1994, two years before his death, he remarked,

During the past year, I have read a number of articles that make my jaw drop. They aren't from our [the computer science] community. They are from the molecular biology community, and I can imagine two ways of riding the coat tails of a much bigger revolution than we have. One way would be to attempt to make computing elements out of biological devices. Now I'm not very comfortable with that because I am one and I feel threatened. I prefer the second course, which is to use biological devices to manufacture non-biological devices: to manufacture only the things that are more familiar to us and are more stable, in the sense that we understand them better. . . . Or to come right to the point, how do we train bacteria to make transistors?[3]

Thus, with such influential and estimable antecedents as these, this book purports to be a nonexhaustive survey of the emerging field of biologically inspired computation, written at the level of a technical generalist. Though somewhat of a catch-all term, "bio-inspired" computation can roughly be defined as: (1) the use of biology as metaphor, inspiration, or enabler for developing in silico algorithms; (2) the construction of new information processing systems that either use biological materials (e.g., cells, enzymes) and/or are modeled on biological processes, even if they're made from conventional materials; and (3) understanding how biological organisms "compute," that is, process information. The term biologically inspired computing, as understood here, will not include the use of computers for analysis or data management in biology, as, for instance, in bioinformatics or computational biology. That said, the very breadth and diversity of the research at the bio/computing interface makes it evident that there already exists a fertile synergy between the two fields.

The topics covered here will include artificial neural networks (chapter 1), evolutionary and genetic algorithms (chapter 2), cellular automata (chapter 3), artificial life (chapter 4), DNA computation (chapter 5), self-assembly (chapter 6), amorphous computing (chapter 7), computer immune systems (chapter 8), biohardware (chapter 9), and the "computational" properties of cells (chapter 10). Chapters 2–4 take up fields that were among the first developed to combine information science and biology, several decades ago. Though much has already been written about them, I believe they belong in this volume from a thematic standpoint. Moreover, some readers may be unfamiliar with the history and developments in these fields and their role as important tools in scientific research.

The survey presented here is not meant to catalogue all the scientific and technological developments to date in the bio-inspired computing field, as that is beyond its scope. However, the book does attempt to cover some of the more salient or representative areas, in addition to giving a distinct flavor of the extraordinary range, variety, and originality of research in this field. Even with these contraints, I had to exclude some projects for lack of space or because, unfortunately, I was not familiar with them.

It's only fitting to acknowledge the help of many people in getting this book written. Contentwise, Hal Abelson, Tom Knight, Phil Kuekes, John Reif, Carter Bancroft, Andy Berlin, Chris Adami, Harley McAdams, Roger Brent, Drew Endy, Erik Winfree, David Fogel, Stephanie Forrest, Jeff Kephart, Tim Gardner, Jim Collins, Laura Landweber, Grzegorz Rosenberg, Jeff Mandula, George Whitesides, and Mike Foster were especially helpful, for either reviewing portions of the draft or teaching me about the field.

I'm also grateful to Ron Weiss, Radhika Nagpal, Gerry Sussman, Dick Lipton, Sonny Maynard, Sri Kumar, Paul Messina, Leila Kari, Anne Condon, Len Adleman, Bob Birge, Jeff Stuart, Devens Gust, Neil Woodbury, Hod Lipson, Andre DeHon, Daniel Mange, Moshe Sipper, Andy Ellington, Rob Knight, Jordan Fiedler, Adrian Stoica, Tom Ray, Tetsuya Higuchi, Charles Taylor, David Jefferson, Harold Morowitz, Ken de Jong, Karl Sims, Ken Musgrove, Tom Ray, John Koza, Jose Munoz, Sharad Malik, Animesh Ray, Andy Tyrell, Paul Werbos, Bud Mishra, Michael Elowitz, Michael Simpson, David Fogel, Xin Yao, Andy Tyrell, William Aspray, Ned Seeman, Christine Keating, Tom Mallouk, Steve Hofmeyer, Bill Wulf, George Spix, David Kahaner, Roberto Piva, and all the members of the BioSPICE Developers Group who responded to my queries.

This book was written during evenings and weekends while I continued to work at my "day job," without the luxury of "time and tranquility," in the words of Einstein, to enable me to delve as deeply into the topic as I would have liked. Family and friends were generally responsible for moral support, advice, or prodding me to write. My thanks go to my uncle Gordon Forbes, Simki Kuznick, the Mandulas, Mary Clutter, Linda Arking, members of the DC Chapter of the Association for Women in Science (AWIS), Freeman Dyson, Lewis Branscomb, Mark Zimmerman, Gwen Suin-Williams, Kathy Rones, Laura Auster, the Schebesta family, the Trotta family, the Valdes family, the Codina family, Harriet Martin, Robin Appel, the Rogers family, Jude Colle, Ann Ladd, the Hillebrands, Karen Ruckman, my mother and sister, and my three cats, Allegra, Cara, and Isis. Special thanks to Mat Burrows for lending me his cottage near the shore in the final stage of the book, and especially to Rita Colwell for her generous and unflagging support of my work, and that of other women in science.

Finally, this book could not have been written without the help of the MIT Press people, including Katherine Innis, Valery Geary, Katherine Almeida, Yasuyo Iguchi, and most especially my editor, Bob Prior, on whose support I depended the most.

Imitation of Life

1

Artificial Neural Networks

There is nothing either good or bad but thinking makes it so.
—Shakespeare, *Hamlet*, II, ii

Since man's earliest efforts to build an electronic calculating machine, scientists and engineers have dreamed of constructing the ultimate artificial brain. Though we may never reach this goal, the first successful attempt to create a computer algorithm that would mimic, albeit in a much simplified way, the brain's remarkably complicated structure and function represented a significant stride forward. These algorithms, known as artificial neural nets, are defined as an interconnected group of information processing units whose functionality is roughly based on the living neuron. As these units "learn" or process information by adapting to a set of training patterns, it is reflected in the strength of their connections.

Neural nets represent a different paradigm for computing than that of conventional digital computers, because their architecture closely parallels that of the brain. (Traditional computers, based on von Neumann's design, were inspired by a model of brain function by incorporating concepts such as input, output, and memory, but reflect this only abstractly in their architecture.) Neural nets are useful for problems where we can't find an algorithmic solution, but can find lots of examples of the behavior we're looking for, or where we need to identify the solution's structure from existing data. In other words, they don't need to be programmed to solve a specific problem; they "learn" by example. They have their roots in a pioneering 1943 paper written by mathematician Walter Pitts and psychiatrist Warren McCullough, "A Logical Calculus of the Ideas Immanent in Nervous Activity." It was the first time anyone had tried to describe the idealized behavior of the brain's network of nerve cells (neurons)—a poorly understood phenomenon at that time—in the language of mathematics and logic.

A Logical Calculus for the Brain

Though the Pitts-McCullough theory had its shortcomings—many physiologists were not happy with its treatment of the neuron as a black box that followed certain mathematical rules for input and output without taking actual physiology into account—many of their ideas were revolutionary and still survive today. They were the first to bring mathematical uniformity, based on logical axioms, to the idea of information processing in the brain. Within this framework they described a network of neurons that cooperated to sense, learn, and store information, in addition to other information processing tasks. They originated the highly sophisticated way of conceptualizing a neuron as an element that sums the electrical signals from many incoming neurons, in addition to the notion that the strength of the synaptic connection between neurons acts as a weighting function whose value determines whether the outgoing signal will excite or inhibit an outgoing electrical nerve impulse. The two researchers also originated the idea that the weighted sum of nerve signals coming into a synapse had a threshold value. According to this neural calculus, if the sum exceeded this value, the outgoing signal would be a one; if not, it would be a zero. This demonstration of the digital nature of neural behavior would come to be a key concept in the theory of artificial neural nets.

The work of psychologist Donald Hebbs also helped shape the field of artificial neural nets. His 1949 book *The Organization of Behavior* put forth the idea that the more active two connected neurons were, the stronger their synaptic connection would become. By extension, the greater the degree of electrical activity throughout a given neural pathway, the more its synaptic connections will be reinforced; this effect, on another level of abstraction, equates to "learning." In other words, according to Hebbs, when we learn something—a child learning to write the letters of an alphabet by repeated practice, for example—we are strengthening the connections in the brain's neural pathways that underlie the behavior.

A New Field Evolves

Artificial neural nets were the first computer algorithms that attempted to model not only the brain's organization, but also its ability to actually learn, based on physiological changes in the organization of neural pathways. The way one actually goes about training a neural net to perform a task is complex; however, it's a vastly simpler process—acting on a vastly simpler network structure—when compared to the biochemistry of learning in a living brain, even as we learn something as rudimentary as a one-syllable sound.

Since its origin in the 1950s, based on Pitts and McCullough's work, the field of artificial neural nets has continued to develop by drawing conceptually from advances in modern neurophysiology, as well as by repeated application at the hands of researchers. For example, a special class of artificial neurons called "perceptrons" were created in the late 1950s. The perceptron was more elaborate and more akin to biological reality than the more schematic, mathematically based Pitts-McCullough neuron, and was defined as a single layer of information processing units that transmitted signals and adapted its interconnecting weights accordingly. In the late 1960s, however, further analysis revealed that the perceptron was unable to carry out certain logic functions involved in more sophisticated learning algorithms. Though these problems were eventually solved, this failing of the perceptron slowed growth in the field and brought it to an eventual intellectual impasse, causing interest in artificial nets as a model for human intelligence to dwindle.

In the early 1980s, however, the field experienced a rebirth, largely prompted by the discovery of "recurrent networks" by Caltech physicist John Hopfield. These are networks where information flows from a connection node back to itself via other nodes, providing all the neurons in the network with complete connectivity and greater resemblance to the biological brain. (In other words, they incorporated self-feedback loops.) This development added greatly to the range of problems neural nets were capable of addressing. Hopfield was also responsible for introducing the idea of "hidden" layers of neurons between input and output layers (figure 1.1). These layers are not connected to the outside and can recode, or provide a representation for the input units. They are more

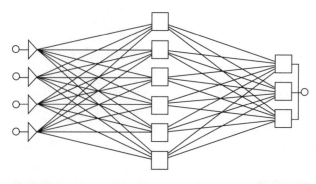

Figure 1.1
Artificial neural nets usually have one or more "hidden" layers between input and output layers.

powerful than single-layer networks, in that they can learn a much wider range of behaviors.

The next big development came in the 1980s with the arrival of the "backpropagation" algorithm. This was a procedure for training neural nets to learn from test cases or "training sets." These are presented to the net one at a time, and the errors between the actual and desired behavior of the network are propagated backward to the hidden layers, enabling them to adjust the strength of their connections accordingly. This method is then iterated to reduce the error to an acceptable one. These two advances were probably responsible for a large resurgence of interest in the field in the late 1980s, and though artificial neural nets never became the much hoped-for means for creating an electronic replica of the human brain, they did give rise to a sophisticated technology that is still widely employed today in many industrial, research, and defense applications, particularly for pattern recognition.

How Artificial Neural Nets Work

The mammalian brain is made up of a huge network of nerve cells or neurons that are specialized to carry messages in the form of electro-chemical signals. In humans, the brain has more than 100 billion neurons that communicate with each other via a massive web of interconnections. These interconnections consist of nerves called dendrites, which carry input into the neuron, whereas other nerves, called axons, are its output channels. At the connection points between the dendrites and the axons—the synapses—the electrical impulses flowing down the axon get transformed into biochemical signals, cross the synaptic gap, and are then re-transformed into electrical signals that travel up the dendrite to the next neuron. The electrical impulse passing though the dendrite is either "excitatory" (promoting action) or "inhibitory" (inhibiting action) in nature. If the difference between the sum of all excitatory and inhibitory impulses reaching the neuron exceeds a given threshold, the neuron will fire an electrical pulse. This pulse, in turn, is itself inhibitory or excitatory in nature (figure 1.2). This represents the *all-or-none* response of the neuron.

Artificial neural net algorithms are based on a highly simplified model of the brain's elaborate network of neural connections. Artificial neural nets have input channels that represent the dendrites, and output channels that mimic the axons. The synapse is modeled by an adjustable weight, located at the juncture between incoming and outgoing channels. A one or zero represents the corresponding excitatory or inhibitory signal that flows out from each connecting point (figure 1.3).

Within the artificial neural net, each connection weight modifies the incoming signal before sending it on by assigning it an appropriate weighted value. Much like what happens in the living brain, all the weighted input signals in the network that flow into that particular

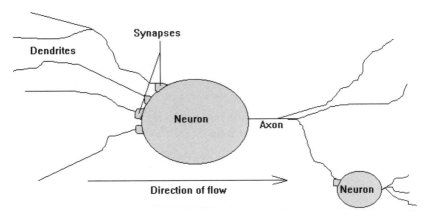

Figure 1.2
Sketch of a living neuron, showing dendrites, axons, and synapses.

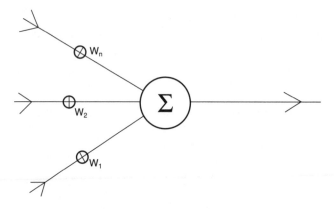

Figure 1.3
Simplified artificial neuron. Each input is multiplied by a weighting factor
(w_1, w_2 . . . w_n), before flowing into the "synapse," where they are summed.

synapse are added together to form a total input signal, which is routed
through something called an "input-output function." This acts on the
signal to form the final output signal. The weights and the input-output
function are what ultimately determine the behavior of the network, and
can be adjusted (figure 1.4).

In order for the artificial neural net to carry out a useful task, one must
connect the neurons in a particular configuration, set the weights, and
choose the input-output functions. The simplest artificial neural net
would consist of a layer of input units connected to a single middle or
"hidden" layer, which is linked to a layer of output units. To initialize
the artificial neural net, whatever raw data is needed to perform the task
is first fed into the input units. The resulting signal received by a neuron
in the hidden layer depends on how the incoming raw data is weighted,
and how it is modified by the input-out function. In the same way, the
signal flowing out of the hidden layer goes through a similar process of
weighting and modification before going on to the subsequent level.

What makes artificial neural net algorithms so valuable is that they
can be taught to perform a particular task, such as recognizing patterns
inherent in an incoming data set. A concrete example may help demys-
tify the process by which artificial neural nets learn. Suppose we want
to train the network to recognize handwritten letters on a display screen
(as many credit card machines do today with the card owner signature).

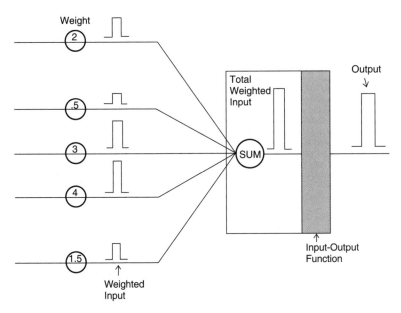

Figure 1.4
Detailed artificial neuron. All the weighted inputs are summed to form a total weighted input, which then passes through a given input-output function for computation of the final output signal.

The artificial neural net would therefore need as many input units as the number of pixels making up the screen, and twenty-six output units for each letter of the alphabet (with any number of units in the hidden layer in between).

To train this network to recognize letters, we first present it with the image of a handwritten letter, for example, "A," and compare the output this input signal produces with the desired output. (The learning task is incremental, as the network gradually learns the desired task). We then calculate the discrepancy or error between the initial output and the one we ultimately want. The error is defined as the square root of the difference between the two outputs. Then we tweak the values of the weights a little in an attempt to better approximate the desired output, continuing this iterative process until the actual output comes closer and closer and finally matches the desired output.

The difficult part, of course, is knowing how to modify the weights to increasingly reduce the error between actual and correct values.

However, in practice, the weight is adjusted by an amount proportional to the rate of change in the error, a quantity that can be calculated as the weight is changed. This quantity is called "the error derivative of the weight," or EW. It is often very hard to calculate. In 1974, Paul Werbos then a doctoral student at Harvard, invented a better way to calculate the EW, a method called backpropagation.

To understand how backpropagation works, first assume all the input-output functions are linear. To find the EW, we must first find the rate of change in the error as a particular unit's signal is changed, called the "error derivative," or EA. For an output unit in the network, this quantity is the difference between the actual output and the desired output. In backpropagation, we start backward from the output layer, computing all the EAs for the hidden layers and input layers. To find the corresponding EW for each weight in each layer, we simply multiply the EA by the signal that enters that weight. When all the weights have been adjusted by the right amount, we can input the same raw data into the network, and the actual output will match up with the desired output. In the handwriting recognition case, we can feed into the network the corresponding impulses from the pixels on the display screen as someone writes an A, and the handwriting recognition device will register "A."

Training algorithms for artificial neural nets come in many varieties, the two most common being *supervised* learning and *unsupervised* learning. In the former, an outside computer program monitors the learning process just as a teacher would do for a student; in unsupervised learning, the network is only presented input data, and the system adjusts its own weights without the benefit of knowing the relationship between the input and final output. To reach a solution, the system groups the input data into special classes, and is ultimately able to obtain a single output correlated with each group. Certain artificial neural net algorithms used to recognize spoken speech patterns are unsupervised, in the sense that they place spoken words into different phonetic classes.

Why Artificial Neural Nets Are Useful

Artificial neural nets have been successfully applied to a large number of problems, which usually fall into one of three classes: recognizing something, inferring something, or putting things into classes. Pattern recognition algorithms are probably the most common. Among these are the automated recognition of handwritten text, spoken words, facial/fingerprint identification, and automatic moving target identification against a static background (for example, the ability to differentiate the image of a moving tank from the road).

Speech production is an example of a classification algorithm. Connected to a speech synthesizer, the artificial neural net is able to classify different sounds, such as vowels, consonants, and even those that separate one word from another—as an infant does when learning to speak—developing increasingly finer-grained classes of sounds until they represent actual intelligible speech. Artificial neural nets are also used for industrial control systems, such as power plants or chemical factories. The network is fed data that represents the system's optimal state of functioning (for example, the ideal temperature, pressure, and vacuum conditions) and continually monitors the system for any deviation from these values. Other applications include artificial neural net software to predict stock market trends (a pattern recognition algorithm), or to process signals while canceling out noise, echoes, and other unwanted parts of the input signal.

Artificial Neural Nets and Digital Computers

The way an artificial neural net processes information is fundamentally different from the way digital desktop computers do—although the latter can be modified to run artificial neural nets. Conventional digital computers are traditionally known as "von Neumann Machines" because they are based on von Neuman's original designs. They essentially work by deductive reasoning. This method is optimal for solving problems whose solutions can be reached by following a formalized, linear, finite series of instructions (algorithm) that the computer's central processing unit (CPU) executes. Computers must be programmed *a priori* with the exact series of steps needed to carry out the algorithm. What's more, the data fed into the program must be precise, containing no ambiguities or errors. Conventional computers are amazingly adept at carrying out what they've been programmed to do, including executing extremely complicated mathematics. They are also remarkably fast and precise. Some digital supercomputers can perform more than a trillion operations per second and are thousands of times faster than a desktop computer.

However, traditional digital computers can only solve problems we already know and understand how to solve. They're ineffective if we're not sure what kind of problem we want to solve, know an algorithm for doing it, or if the data we have to work with is vague. But if we can point to a number of examples of the kind of solution we require, or if we simply want to find a pattern in a mass of disorganized data, artificial neural nets are the best method.

In contrast to digital computers, artificial neural nets work by inductive reasoning. Give them input data and the desired solution, and the network itself constructs the proper weightings for getting from one to the other. This is what is meant by saying that the artificial neural net is "trained" from experience—the initial network is built and then presented with many examples of the desired type of historical cause and effect events. The artificial neural net then iteratively shapes itself to build an internal representation of the governing rules at play.

Once the network is trained, it can be fed raw input data and produce the desired solution on its own—analogous to the way the brain functions in the learning process. Unlike the digital computer, where computation is centralized, serial, and synchronized, in artificial neural nets, computation is collective, parallel, and unsynchronized. They tend to be much slower at this process than digital computers—artificial neural net operations are measured in thousandths of a second, whereas digital computers can function at up to teraops, or 10^{12} operations per second rates at present.

Artificial Neural Nets and Artificial Intelligence

Though the initial impetus for developing artificial neural nets may have been a desire to create an artificial brain, in the years since Pitts and McCullough's work, research in the general area of machine learning has grown so specialized and so diverse that some of the algorithms bear little resemblance to others. Such is the case of artificial neural nets and artificial intelligence (AI), although ultimately each represents a different approach to the long-standing quest to make computers more and more humanlike in their abilities.

Artificial intelligence, like artificial neural nets, consists of computer algorithms that mimic human intelligence. They are typically used to carry out tasks such as learning, game playing, natural language processing, and computer vision. Generally speaking, artificial intelligence differs from artificial neural nets in the level of human intervention it requires. With an AI algorithm, all the information needed for a solution must preprogrammed into a database, whereas artificial neural nets learn on their own. AI is based on the principles of deductive reasoning, whereas neural nets are inductive. This means that with AI, each new situation the system encounters may require another programmed rule. For example, when AI is used to program the behavior of a robot, all the desired behavior patterns must be worked out and programmed a priori—the robot can't adapt its behavior to changes in the environment. Consequently, AI programs can become quite large and unweildy in their attempt to address a wide range of different situations.

Artificial neural nets, on the other hand, automatically construct associations or relationships between parts of the network according to the results of known situations, adjusting to each new situation and eventually generalizing their behavior by correctly guessing the output for inputs never seen before. The disadvantage of artificial neural nets, however, is that they cannot be programmed to do a specific task, like adding numbers. The sets of examples or "training sets" of data the network must be fed in order to bring it closer to the desired solution must be chosen very carefully; otherwise, valuable time is wasted—or worse, the network doesn't do what it is supposed to do.

Popular culture has conditioned us to expect the future to be populated with robots containing computer programs that will make them look and act like humans. Although this is a quixotic goal that we may never reach—and some would question whether or not it's even desirable—as more of the brain's remarkable complexity is deciphered and understood, it will likely inspire many new technological ideas, equally as impressive as these.

2

Evolutionary Algorithms

Accidents will occur in the best regulated families.
—Charles Dickens, *David Copperfield*

Evolutionary algorithms, as the adjective implies, take their inspiration from the macro principles of biology rather than its micro principles, as artificial neural nets do. Indeed, in both form and function, they mirror the dynamics of a theory whose discovery by Charles Darwin in the mid 1800s constitutes one of man's greatest contributions to science.

Evolution as Metaphor

When Darwin's *The Origin of the Species* was published in 1859, biological ignorance was widespread in the world. Most people believed life on earth had originated by heavenly decree; little or nothing was known of the principles of heredity, fertilization, or the development of the mature animal from an embryo. Darwin's theory of natural selection—a radical departure from currently accepted beliefs—was so sweeping in scope that it was able to account for all these phenomena, in addition to the wide variation in the earth's species and the complex inter-relationships among all living creatures. It completely revolutionized nineteenth century natural science revealing a logical and evincible mechanism, called natural selection or survival of the fittest, by which all plants and animals had slowly evolved from earlier forms. The empirical evidence supporting Darwin's theory, as manifest in the fossil record, was so compelling, as to make his ideas almost irrefutable. What Isaac Newton had done for the physical sciences two centuries earlier, Darwin did for biology.

As explained in *Origin*, the driving force behind evolution is a push to make each successive generation of organisms better able to survive, adapt, and prosper in its environment so it can reproduce and continue to maintain the species. As a rule, each generation of a species usually produces many more offspring than needed to maintain it; however, the size of its population tends to remain constant because disease, competition, and other erosive forces eliminate the organisms that are weaker or less well adapted to their environment. The notion of *fitness* provides a criterion for how well an individual organism can adapt to its environment, and will ultimately determine whether it survives or not. The characteristics that cause an individual to be fit, such as a camouflaged coat for hiding from predators or keener eyesight to spot food, are largely passed on to its progeny; eventually the species as a whole will inherit these characteristics, becoming more robust and resilient as a result. Fitness provides a measure for the breeding success of any two parents in a given population in a given generation, because the fittest individual is ultimately the one that will contribute the most offspring to the next generation.

Digital Darwinism

The dynamical principles underlying Darwin's concept of evolution have been co-opted by computer scientists to provide the basis for a class of algorithms that are extremely well suited for solving some of the most demanding problems in computation. Just as the brain and its web of neural impulses inspired the creation of artificial neural networks, biology has also been the impetus for the development of an ingenious and highly efficient method for computer optimization. These computational equivalents of natural selection, called evolutionary algorithms, act by successively improving a set or generation of candidate solutions to a given problem, using as a criterion how fit or adept they are at solving the problem.

Evolutionary algorithms have many applications. One use is for searching large databases of possible solutions to a problem to find the best one. For example, bioinformatics, a field that uses mathematical, statistical, and computational methods to solve biological problems involving DNA, amino acid sequences, and other related information often uses evolutionary algorithms. When trying to find the structure of a large biomolecule, given the various configurations the molecule could take under a given set of conditions, say, of acidity, temperature, or electrical charge polarization, these algorithms can analyze and predict possible shapes or geometries the molecule will assume, until finding the optimum one, even providing a genetic sequence when DNA is involved. To do this requires a process called sequence analysis, which conducts a search among vast numbers of possible sequences for DNA to find the right one. The fact that evolutionary algorithms can be highly parallel— that is, they can carry out many identical operations simultaneously— makes solving these sorts of problems more tractable. What's more, the inherent notion of *fitness* provides explicit criteria for identifying the optimum answer.

Evolutionary algorithms also work extremely well with problems that must accomodate and adapt to changes in the environment during the computer run time. For instance, a robot might have an evolutionary software program embedded inside that enables it to adapt to changing spatial coordinates, so as to maneuver around a room without colliding into surrounding objects, while receiving constant input from sensors on its position.

Genetic Algorithms and Genetic Programs

As expected, the functional principles underlying evolutionary algorithms are intrinsically Darwinian in nature: those members of a given population (or set of possible solutions to a problem) that are the fittest and can best adapt to their environment will be the winners in the game. The principles were first developed four decades ago as a tool for optimization in engineering problems. In the 1960s, two German scientists, Ingo Rechenberg and Hans-Paul Schwefel, came up with a method of programming called "evolution strategies," which they used to help find the optimal aeronautical design, such as the shape of airplane wings. Around the same time, completely independently, Lawrence Fogel in the United States developed a method of computational problem solving he termed "evolutionary programming," which also drew on the principles underlying natural selection. Without going into detail, the latter algorithms differed somewhat from the former in the way they represented candidate solutions and in their formulation of the mechanisms used in producing the next generation of "fitter" solutions.

A few years later, John Holland at the University of Michigan—where he currently holds faculty positions in the Psychology, Computer Science, and Electrical Engineering Departments—invented genetic algorithms (GAs), today considered a subset of the larger category of evolutionary algorithms. GAs were different from their precursors, evolution strategies and programming. Although all shared the same fundamental goal of arriving at the most favorable solution to a problem via iterated natural selection, the earlier algorithms used methods for reproduction and fitness that were limited in scope and didn't exploit all the diversity and subtlety of Darwin's evolutionary principles. Holland's algorithms, by contrast, were deeper and more comprehensive. He had attempted a full-fledged translation of the actual empirical mechanisms of biological adaptation into the language of computer science in his first book on the subject, *Adaptation in Natural and Artificial Systems*. His purpose was not simply to borrow ad hoc from biology to solve a specific problem, but to essentially marry the two fields and in doing so, produce a completely new subfield in computer science.

According to his scheme, the "chromosomes" of natural selection (i.e., the strings of DNA that provide the blueprint for a biological organism and form the basis for heredity) are converted into strings of ones and zeros (bits), although genetic algorithms don't have to use this representation. From these, one could construct a population of chromosomes representing the candidate solutions to a problem that would reproduce according to the rules of natural selection to yield a new generation of chromosomes, better adapted to thrive in their "environment," that is, to solve the problem at hand.

In the same vein, Holland invented several "bio-operators," called *crossover*, *mutation*, and *inversion*, to accomplish this task. These were applied to each individual in a generation allowed to reproduce—usually those with the fittest genes.

The *crossover* operator, as its name implies, interchanges subsequences of two chromosomes to produce two offspring. For example, suppose the two parent chromosomes consist of the following bit strings: A = 011100 and B = 110101. The crossover operator would cut each string in half, transposing such that the second half of B replaces the first half of A, to form 101100. Similarly, B, after crossover, becomes 100101. Crossover approximates the biological process of genetic recombination between two organisms, each of which have only one set of unpaired chromosomes (called *haploid* organisms, as opposed to *diploid*, which are those that have two sets of chromosomes, one from the father and one from the mother, present in one cell).

Another bio-operator, *mutation*, randomly flips some of the bits in any given chromosomal string belonging to either or both of the two parents to create a new offspring. For example, mutating the original A might make it 111000, by flipping the first and fourth bits, as chosen at random.

The *inversion* bio-operator produces offspring where the order of a segment of the parent chromosome is reversed—for example, changing the original B to 011101, by reversing the order of the first three bits, 110, to 011.

Holland's creation of algorithms incorporating genetic operators inaugurated an important new phase in the development of evolutionary algorithms. His work provided the underpinnings for a coherent theoretical foundation for the field, and in time gave rise to several different varieties of evolutionary algorithms.

One of these was genetic programming, developed by Stanford's John Koza, who sought to combine genetic algorithms with the basic concept of AI. Genetic programming uses evolution-inspired techniques to produce not just the fittest *solution* to a problem, but an *entire optimized computer program*. Instead of a population of bit strings, it uses program fragments and subjects them to operations such as crossover or mutation. Koza's genetic programming enabled researchers to come up with a set of solution programs that were conventional computer programs in the sense that the computer could automatically run them.

Genetic programming is not expressed in the form of lines of code, analogous to evolutionary algorithm's use of bit strings; rather it is represented in the form of a "parse tree," or a tree whose branches subdivide at nodes. When a genetic operator such as crossover is applied to the genetic program, two subtrees or branches are exchanged instead of two bit strings. At each node sits a particular function or control structure, whose arguments are functions, variables, or constants.

For example, if genetic programming were to be used to operate a robot, a node might return a value obtained from input sensors or an instruction for the robot to follow. Koza's algorithms seem particularly suited for solving problems in computational molecular biology (for example, defining a biochemical signaling network that a particular gene forms when producing a protein) or for designing analog computer circuits. Chapter 8 covers the use of evolutionary algorithms for circuit design—evolvable hardware—in more depth.

Survival of the Fittest Solution

Seeing how evolutionary algorithms are applied to a specific problem may help enable greater appreciation of the method's power and sophtication. For example, suppose a robot is programmed to get from point A to B in five minutes. However, in order to accomplish this, it must pass through a maze (figure 2.1). (Collision avoidance software actually uses evolutionary algorithms to enable robots to find their way around obstacles.) To make its way through the maze, the robot has only four allowed motions: Back, Forward, Right, or Left. The algorithm assigns a repre-

Figure 2.1
The robot must navigate its way through the maze. Allowed motions are backward = 000; forward = 100; left = 001; and right = 101.

sentation for each direction, the simplest one being binary logic, by assigning a bit string of ones and zeros to each motion—for example, B = 000, F = 100, R = 101, and L = 001. As the algorithm uses number of steps as its main variable, in order to put the five-minute constraint into program, that must also be expressed in terms of steps.

Assuming that the robot travels at a fixed speed of one step every three seconds, without backtracking, in five minutes it can go a hundred steps. Hence, the robot is allowed exactly one hundred steps to go through the maze, no more, no fewer. Obviously, there are many paths it can take—some of which are better than others—and all have steps made up of some sequence of B, F, R, and L moves. Each possible move is translated into a hundred three-bit strings to generate a population of candidate paths. The bit strings act as the chromosomes, which will reproduce while competing with each other to produce a fitter path, one that takes exactly a hundred steps to go from A to B.

To calculate the fitness of any given path, we let the robot follow it through the maze and calculate its position after a hundred steps. The fittest path will be the one that, when done, has zero steps between its stopping point and B. If there is ultimately no path that enables the robot to reach B in a hundred steps, then the path that gets closest to B with smallest number of steps remaining before reaching it is the fittest path.

To start the algorithm, we first generate a reasonable number of completely random paths from A to B to form the first population. Then we calculate the fitness of each path, using the above definition. We then select a pair of paths (call them the parents), and with a probability p proportional to the fitness value of each, use the crossover operator to produce two new offspring. If no crossover is carried out (with probability $1 - p$ that this happens), then the two offspring remain the same as their parents. Next, we apply the mutation operator to the two offspring to produce two members of a new generation. To complete the algorithm, we repeat all the steps starting with calculating fitness values, to produce a completely new "second generation" population. Each cycle of operations represents one generation of the algorithm. A standard evolutionary algorithm would take anywhere from 200 to 500 generations to find the best solution. In the case of our robot, after a certain number of cycles, we can test the fitness of the resulting paths, select the fittest one, and then re-translate this back into a series of B, F, R, and L movements that corresponds to the path that takes it through the maze in a hundred steps.

Evolutionary algorithms work well with a remarkable range of problems. They are especially suited for large-scale optimization problems in engineering, control theory, and many similar areas. Despite their successes, however, some users complain that these applications take too many generations to find the solution—a task that is both time and resource consuming on a computer—and also produce a final configuration that is not always straightforward.

Examples of problems that evolutionary algorithms have successfully solved include job-shop scheduling on a factory floor; predicting the particular configuration for protein folding; machine learning, as in the case of the robot "learning" to go through the maze; setting the values for weights in an artificial neural network; or modeling host-parasite behavior over time in ecological problems. Their remarkable versatility also extends to the realm of the social sciences, where they can be used for modeling the evolution of economic markets, or for studying aspects of social systems that evolve over time, such as the evolution of co-operation and communication between large numbers of individuals in a community.

Gameplay

One fascinating application of evolutionary algorithms was devised by David Fogel, CEO of Natural Selection, Inc. in La Jolla, California, and Kumar Chellapilla, then a doctoral student at University of California San Diego. The two performed an experiment using the method to evolve a population of artificial neural nets that represented different strategies for a computer engaged in a game of checkers with a human adversary. The algorithm enabled the computer to adapt to new circumstances that arose during the game without the need for human intervention, something IBM's chess-playing Deep Blue computer could not do. Deep Blue, though able to beat Russian Gary Kasparov, the reigning world chess champion in 1997, could only draw from a fixed repertory of specific, pre-programmed moves. Faced with a particular set of moves, it would carry out the same game every time, with no variation or "learning."

In Fogel and Chellapilla's method, the starting population was composed of fifteen randomly chosen strategies for play, each represented by an artificial neural network. These strategies then competed among themselves, starting from random initializations, obtaining points for each win, loss, or tie in a series of games. The total score gave a measure of the strategy's *fitness*; those with highest scores became parents for the next generation of game plans. Changing the connection weight values of the parent nets created offspring.

After 250 generations, the final evolved neural net was tested in a game against human players on an Internet site, without telling the humans their opponent was a computer program. The evolved programs were able to win against a large percentage of experts (figure 2.2). Its

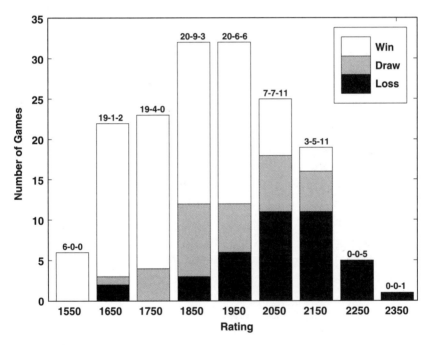

Figure 2.2
The performance of the best evolved neural network after 230 generations, played over 100 games against human opponents on an Internet checkers site. The histogram indicates the rating of the opponent and the associated performance against the opponents with that rating. Ratings are binned in units of 100. The numbers above each bar indicate the number of wins, draws, and losses, respectively. (Graph courtesy of David Fogel, Natural Selection, Inc.)

inventors had only provided the positions of the pieces on the board and the piece differential; the computer was able to learn everything else it needed merely by playing the game.

Evolution for Art's Sake

Visual artists and composers have recently turned to evolutionary algorithms for producing "evolved" works of art and music. Evolutionary art, or evo-art as it's called, probably resulted from seeds planted by biologist Richard Dawkins some years ago. Dawkins experimented on his computer with the creation of two-dimensional shapes called Biomorphs, combining simple rules of evolution to grow visual images. Most evo-artists combine educational backgrounds in both computer science and art, which enables them to write their own evolutionary programs and design their own images.

Their artistic programs are based on evolutionary algorithms, which start by generating a random population of bit strings or instruction networks, although all the artist sees are their visual representations. The artist then assigns each member of the population a score, similar to a fitness score. In this case the score is arbitrarily assigned according to the aesthetic vision of the artist, rather than a logical methodology. The algorithm combines the scored parent images to produce offspring, creating generation after generation of visual images (figure 2.3).

The fact that evolutionary algorithms have also been used to compose music gives an indication of their extraordinary versatility. To use the algorithms in this way, the act of musical composition must be cast as a search problem, with the solution space being the set of all possible musical compositions. The evolutionary algorithm searches through these solutions to find the most harmonious composition.

The problem is set up as a series of three separate modules—the composer, ear, and arranger modules—with each module applying its own notion of fitness. The composer module starts with the output of a random music generator and produces some sort of musical sounds. The ear module then acts as a filter that identifies these compositions as either good or bad sounding, with the notions of "good" or "bad" reflecting the musical taste of the creator. In the ear module, those parent pieces that sound best reproduce. The method uses crossover and mutation

Figure 2.3
Example of evo-art. This image, "Untilled," was generated by Ken Musgrove of Pandromeda, Inc. using genetic algorithms. (Image courtesy of Ken Musgrove)

operators on this population to produce a set of offspring with successively higher fitness (and hopefully musical) values. The third module, the arranger module, then works with these generations to create more ordered compositions that are themselves evaluated, combined, and recombined to produce a more melodious final rhythm.

Though some evo-composers complain that the original search space is so big that convergence to a final solution takes too long to produce, many observers claim to be surprised at the amazingly good quality of music that results from this method.

Evolutionary algorithms show how biological principles can be applied to computation to enable each successive population of hypothetical solutions to a problem to become increasingly fitter and more optimum over time. Similarly, artificial neural nets also enable an algorithm to get trained to be iteratively smarter and better at solving a problem, over each run of the artificial neural net cycle. Paradoxically, in both cases, although each cycle or generation run on the computer takes us closer to our goal of solving a problem, the methods produce increasingly more complicated results rather than simpler ones. Indeed, the lack of straightforwardness of, and the difficulty retracing the development of the final answer is one of the complaints scientists have leveled against both evolutionary algorithms and artificial neural nets.

In the final analysis, whatever their limitations these algorithms retain the efficiency and adaptability of natural processes. As a child develops, learning to speak, read, write, and reason, the number of synaptic connections grow to form a much more complex, yet more highly proficient human neural network. Likewise, living organisms on earth have evolved over millions of years from the lowly bacteria to the highly sophisticated human being. Each individual of a given generation that out competes others does so because it is fitter and better able to adapt, and as a consequence the species eventually develops new capabilities and a more highly organized structure. More highly developed capabilities equate to higher order complexity in what is inevitably the one-way path of evolution.

3

Cellular Automata

The machine does not isolate man from the great problems of Nature, but plunges him more deeply into them.
—Antoine de Saint-Exupéry, *Wind, Sand and Stars*

Scientists have historically approached the study of living phenomena and that of the physical world in two very different ways. Their investigation of biology has been largely empirical and analytic in nature, whereas disciplines such as physics and chemistry, relying just as heavily on experimental results, have lent themselves more readily to the formulation of fundamental laws expressed in mathematical form. Possibly because the evolutionary process living organisms undergo is so complex, and the forms they assume so manifold, scientists have yet to arrive at a set of fundamental laws for biology as they have done for example, in physics with Newtonian mechanics or quantum mechanics.

Von Neumann's Automaton

Von Neumann, however, being a mathematician, believed that by reducing living phenomena down to their barest abstract essences—basically seeing biological parts as black boxes that react to simple but well defined stimuli—it was possible to discern their underlying logic. Without ever losing sight of the enormous complexity of biological organisms, he felt that this method would enable one to see life as a concatenation of dynamical events and interactions carried out by a highly organized system of elements, such as genes or proteins. From this viewpoint, information formed the basis for natural processes to occur, and through its creation, duplication, storage, and processing, enabled the organism to grow (ontogeny), and reproduce.

In his 1948 lecture at the Hixon Symposium, "The General and Logical Theory of Automata," von Neumann presented his idea for an abstact theory of both natural and artificial automata based not on empiricism, but on pure mathematical logic. Although he did not explicitly define what he meant by "automata," we can infer that he generally meant any system that processes information as part of a self-regulating mechanism.[1] In his view, examples of automata included the human nervous system, the computer, and possibly even systems such as radar or telephone communications. This abstract, purely logical design for an automaton, he felt, might help shed light on the logic inherent in the functioning of the human nervous system and also suggest ways to improve the organization and efficiency of computer design.

Natural organisms are, as a rule, much more complicated and subtle, and therefore much less well understood in detail, than are artificial automata. Nevertheless, some regularities which we observe in the organization of the former may be quite instructive in our thinking and planning of the latter; and conversely, a good deal of our experiences and difficulties with our artificial automata can be to some extent projected on our interpretations of natural organisms.[2]

His Hixon lecture also contained a discussion of the common properties of both biological and artificial automata, such as their mode of processing information (electrical switching in both the case of the brain and the computer), the structure and function of the various components that make up automata, their overall organization, degree of complexity, method for dealing with errors, and self-reproduction. By analyzing these common elements in the abstract, through purely logical and axiomatic means, and treating component parts as black boxes, von Neumann was able to place both computer and living organism on the same plane, as it were, and make comparisons between equivalent elements, providing unusual insight into both systems. For example, he observed that natural systems have a highly superior way of dealing with "faults" (e.g., disease, injury, or malfunction in general) through self-healing mechanisms, and that they typically have organizational structures and use materials that are also far superior when compared to those of artificial automata.

The Turing Machine

In addition to the early work of Pitts and McCullough on neural logic, the ideas of Alan Turing contributed substantially to von Neumann's theory of automata. While still an undergraduate at Kings College, Cambridge in the 1930s, British mathematician Alan Turing had formulated a concept, later known as the Turing machine, that today is considered a fundamental cornerstone in the modern theory of computation. The Turing machine was an abstract device that more or less resembled the head of a cassette player. As this tape head passed over an infinitely long ticker tape, it had the ability to read instructions written on the tape, in addition to being able to erase them and write new ones (figure 3.1). According to the instructions it read, the Turing machine was also capable of changing its "state," in accordance with a predetermined finite set of internal rules contained in the tape head. It could assume any one of a finite but very large number of states, with all the specifications for each state laid out by instructions on the tape.

This ingenious concept enabled Turing to prove a very important hypothesis: that the description of any Turing machine in the world could be provided to any other machine as information on the tape—even if it were extremely detailed, the description could ultimately be completely

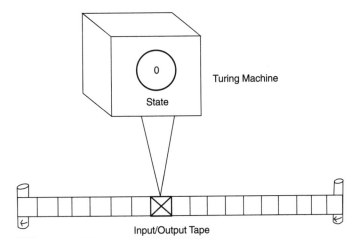

Figure 3.1
Depiction of a Turing machine.

specified—so that the second Turing machine could completely emulate the first, or any other Turing machine. In fact, the ultimate Turing machine would be a universal computer in the sense that, given enough time and the correct instructions, it could simulate any other machine in the world whose behavior could be completely specified by a set of instructions.

As a result, this universal Turing machine could perform any calculation in the world, no matter how complex. This is known as the Church-Turing hypothesis. Its importance comes from the fact that it gives a highly concise theoretical definition of what ultimately can be computed and what cannot.

The Logic of Biological Systems

From a purely logical standpoint, Church-Turing could even apply to a Turing machine that emulates the biological functions of a living organism, insofar as many biological systems can be modeled or described by mathematics, albeit in much simplified form. As we've seen, neurons in the brain transmit electrical impulses through a complex network of connections, forming the physiological basis for much of our mental activity—in particular acquiring, processing, and storing information. If the neuron's input and output electrical signals can be approximated by mathematical formulae, at least in principle, any Turing machine could compute anything the brain could.

Pitts and McCullough argued strongly for this, asserting the universality of their neural networks to perform any information processing task. They felt all Turing machines were equivalent, whether biological or artificial, and in theory could perform the same information processing tasks, no matter how complex or lengthy.

Von Neumann agreed with their contention, countering the general belief of the time that the functions of the human nervous system were so complicated that no ordinary mechanism could possibly describe or model them:

The McCullough-Pitts theory puts an end to this. It proves that anything that can be exhaustively and unambiguously described, anything that can be completely and unambiguously put into words, is ipso facto realizable by a suitable finite neural network.[3]

Parenthetically, proponents of artificial intelligence, decades later, have invoked this argument to demonstrate that computers are indeed capable of simulating a human brain and acquiring "intelligence."

The Kinematic Self-Replicating Automaton

Von Neumann's automaton was also supposed to possess the ability to reproduce. To explore the ramifications of this concept, von Neumann turned again to the Turing machine, coupled with the inherent logic of biological systems that he had outlined for his theory of automata.

He wanted to devise a Turing machine that could not only read and execute instructions, but also make copies of itself. This machine would be capable of movement, and for purposes of reproduction, could construct both a tape and a copying machine out of rigid parts. The copying machine should be able to read the tape and build another automaton, as specified by the tape's instructions. By paring down the concept of self-replication to its essential rules of logic, he created a model for an abstract "creature" that could self-replicate, known as the kinematic self-replicating automaton. Von Neumann's creature had to satisfy the following criteria:[4]

1. It must be a living system that contains a complete description of itself.
2. To avoid an internal self-contradiction, the living system does not try to include a description of the description in the description.
3. In order to avoid contradictions, the description must serve a dual function: (a) it is a coded description of the *rest* of the system; and (b) it is a working model of the entire system, that need not be decoded.
4. The living system contains a supervisory unit that "knows" about the dual role of the description and makes sure that both roles are used during the process of reproduction.
5. Another part of the system, the universal constructor, can build any of a large class of objects—including a system identical to the first one—if it is supplied with the proper materials and is given the correct instructions.
6. Reproduction takes place when the supervisory unit tells the universal constructor to build a new copy of the system, which also contains a copy of the description of the system.

Von Neumann envisioned this machine as being constructed of switches and other moveable parts analogous to organs and appendages in living organisms (figure 3.2). His creature had a mechanical hand that moved about as a robot arm, another hand capable of cutting, and a third appendage that could join two parts together. It also possessed a sensing element that could recognize outside objects and convey this sensory information to the machine's "brain." It had girders that made up the machine's frame and provided a place for storing information.

This automaton lived in a large reservoir full of machine parts, so it could continually pull different parts from the reservoir and use them to construct other machines, according to the instructions fed to it by its control unit. To self-reproduce, the automaton had to read the instructions, copy them, and then store the original instructions, while feeding the copied instructions to the unit that gathered the parts from the reservoir, and then from them, construct a new machine, piece by piece. When the new machine was completely built, it was also given a fabricating unit, a copying unit, and a control unit, just like its parent. The parent automaton then gave a copy of the original stored reproduction instructions to its new offspring, to endow the latter with the ability to repro-

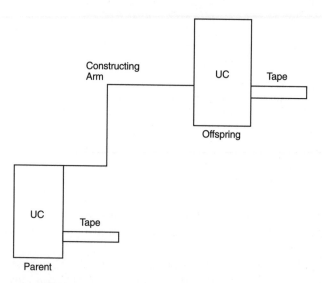

Figure 3.2
Schematic diagram of von Neumann's kinematic self-replicating automaton. (Courtesy of Moshe Sipper, Swiss Federal Institute of Technology, in Lausanne, Switzerland)

duce. So not only was his original machine capable of making copies of itself, but it also gave its "children" the instructions for reproducing themselves.

This last step bore the mark of von Neumann's brilliance and foresight. By basing his machine on the essential logic needed to ensure self-replication (his system contained a description of itself that was copied and given to its progeny), he posited the existence of a mechanism that prefigured the role of DNA within living organisms—which hadn't even been discovered yet! Though von Neumann knew about gene function and remarked that it was clear that the copied instructions given to the child automaton mimicked the function of genetic material in living organisms, the mechanism of DNA that encodes instructions for the development of the adult organism was unknown at the time. Only several years later, in 1953, were molecular biologists James Watson and Francis Crick able to empirically elucidate its function and structure using x-ray crystallography. Now, after many decades, the fact that the concepts of chromosomes, genes, and cloning are so familiar to us that they form part of our common parlance, may make it hard to fully appreciate the extent of von Neumann's prescience. Or as physicist Freeman Dyson declared, "So far as we know, the basic design of every microorganism larger than a virus is precisely as von Neumann said it should be."[5]

Cellular Automata

One problem with von Neumann's kinematic model, however, was that although it functioned well enough on an abstract level, serving to demonstrate that self-reproduction could be modeled solely by the rules of logic, it proved much too complicated to actually build. As a way out of this dilemma, von Neumann's friend and fellow mathematician, Stanislaw Ulam, suggested he simplify his original concept and reformulate the automaton as an "organism" in the form of an infinite grid made out of "cells," like a giant checkerboard. Each cell would act as a finite state machine—much like a Turing machine. It had a finite number of different states it could assume, according to its internal instructions—whose state would change in discrete steps according to (1) an internal set of rules; and (2) the states of its four neighboring cells.

This model roughly corresponded to the behavior of his machine parts. The localized behavior of each cell as it evolved in time eventually caused globalized behavior or a pattern to emerge over the entire checkerboard. Thus it could represent a dynamical system in which both space and time are made up of discrete states. This much simplified creature of pure logic became known as a cellular automaton. Using this new cellular model as a basis, von Neumann then devised another replicating machine made up of two hundred thousand individual cells, shaped like a box with a long tail. Each cell's behavior was determined by rule tables and could enter into any one of twenty-nine finite states, with the totality of the cells' actions over time constituting the organism's behavior. According to a complicated set of rules, this creature could also self-reproduce, just as von Neumann's earlier creature had.

The Game of Life and Beyond

Over the years, the popularity of cellular automata as a method for modeling the behavior of dynamical systems has waxed and waned. In the hands of some researchers, it took the form of games, and in others, it became a window into the nature of complex systems. It even became the basis for a computer, the Cellular Automata Machine (CAM), where each cell was represented by a multiprocessor that could only exchange data with its immediate neighbors, as outlined by von Neumann's original rules. Though the computations were relatively simple at the processor level, in the aggregate, the CAM was capable of modeling highly complex behavior in various kinds of systems.

One interesting development involving cellular automata was the Game of Life, invented by Cambridge University mathematician John Horton Conway, in the late 1960s. Conway's cellular automata was much simpler than von Neumann's in that Conway's only had two states: "alive" or "dead." Conway devised a very simple set of three rules that would determine the survival, birth, or death of a cell. He then applied these rules to a system of cells resembling a giant checkerboard, to see what patterns might evolve from simple initial configurations. He was interested in seeing if patterns could grow without limits, and if they could self-reproduce. He was also interested in seeing if his Game of Life could perform universal computation. After a number of years, when the game, or versions of it, had been played by many researchers, including one at MIT who was the first to run Game of Life configurations on an early computer, Conway felt his goal had been fully accomplished: von Neumann's ideas could be realized in a much simpler form, and that the Game of Life could support universal computation. Conway's invention also served as one of the earliest attempts to simulate something resembling life on a computer.

In the final analysis, however, the proliferating patterns produced on the Game of Life board, which often resembled microorganisms in form, may have been the most lifelike aspect of the process. To think that localized rules played out on a grid could represent fundamental biological principles, albeit in much simplified form, seems to belie the daunting complexity of both micro principles of biology, such as genetics, and its macro principles, such as evolution of the fittest. Although von Neumann did believe that one could ultimately analyze the logic underlying certain biological processes, such as neural behavior, he qualified his statement very carefully by positing first a highly abstracted model, pared down to barest essentials, and second, by turning biological parts into simplified black boxes. He seemed ever aware that natural systems were vastly more complicated and nuanced than any system human beings could imagine constructing, and wanted to avoid a facile reductionism.

Cellular automata might have eventually passed into obsolescence, if several scientists hadn't been intrigued enough to continue to develop and refine them. In the 1970s, computer scientist Tommaso Toffoli began to question what cellular automata were truly good for. Were they just mathematical curiosities, or did they have some actual connection with the physical world? Toffoli decided that cellular automata would be extremely useful for representing certain dynamical systems in the world, where localized rules give rise to globalized patterns of behavior. In his view, they could reveal quite a lot about the workings of complex natural systems, such as the evolution of patterns on shells or the spread of forest fires. He also felt that by using cellular automata, he could show how simple local rules could result in phenomena that were self-reproducing and self-sustaining—behavior that the laws of physics could not completely describe or predict. He was also able to prove that the behavior of cellular automata was reversible, that is, that it could be run backwards to its original configuration. Toffoli and a colleague, Norman

Margolis, were responsible for the construction of the first CAM, followed by a series of similar machines specialized for running cellular automata models.

This was followed by physicist Stephen Wolfram's work in the 1980s, which further revealed the theoretical connection between cellular automata and complex systems. Wolfram's thesis was that although natural phenomena may appear to be following extremely complex principles, or even none at all, they are actually governed by a hidden set of simple localized rules. He believed that cellular automata might provide a pathway to discovering some of life's hidden organizing principles, and the computer could become a testbed for simulating the principles governing living organisms and ultimately the entire universe. Though these efforts were significant in and of themselves, and helped to more fully develop the fields of complexity theory, fractals, and artifical life, in the views of most scientists, they fell short of revealing a new way to look at laws of the universe, based on computers, and lacked the momentum to move the field of cellular automata fully into the mainstream of scientific research.

Are Cellular Automata Really Useful?

Though never becoming the *ne plus ultra* of computer modeling tools for studying scientific phenomena as some advocates hoped they might, cellular automata are still valuable as a simple model for physical systems in which globalized behavior can be completely specified in terms of universal rules and localized relations. By building the appropriate rule tables into a cellular automaton, one can simulate certain kinds of collective behavior, involving movement through space or even dispersion. For example, when using cellular automata to formulate the behavior of a particular physical phenomenon (for example, the fluid flow of water molecules), space could be represented by a uniform grid, with each cell containing a few bits of data, such as position, velocity, and so on, the dynamics of the system would be expressed in a universal look-up table. Time would advance in discrete, measurable steps, and each cell would compute its subsequent state according to the rules in the table and the states of its nearest neighbors. Hence, the system's laws would be local and uniformly applied.

Such an approach has typically been used, with varying degrees of success, to model simplified dynamical systems in such various fields as communications, computation, growth, reproduction, competition, and evolution. The model behavior ranges from the collective motion of fluids governed by the Navier-Stokes equations, to forest fires, where the forest consists of a fixed array in which each cell represents an area of the land surface covered by trees, and rules are established to determine how the fire spreads from one tree to another across the landscape.

In the final analysis, the merit of cellular automata as a scientific methodology is still open to debate. Some computer modelers find cellular automata too fine-grained to work with, declaring that it would ultimately be easier to build a complex computer that simulates a hundred thousand cellular automata than to build a hundred thousand automata to simulate the physical world. Others, such as mathematician John Casti of the Santa Fe Institute, find them to be not only versatile, but also capable of representing important aspects of living things, such as DNA structure or skin patterning, which more linear models could not do. Still others prefer simple differential equations to cellular automata, as the former are easier to use and more accurate in modeling general phenomenological aspects of the world, such as the diffusion of heat or a wave in space. (Both cellular automata and differential equations share the common behavior that, as they evolve in time, they describe at each step a different finite state of the system.)

Cellular automata, however, quite unlike differential equations, are based on the principle that global behavior can result from simple, localized rules. In addition to cellular automata, there are other formulations such as complex adaptive systems analysis or chaos theory that are also based on this important idea. All of them have produced meaningful scientific results, and enabled insights into the properties of collective behavior, such as the growth of mob behavior in a crowd, or evolving patterns in biological pigmentation, that probably would not have been possible with more traditional approaches. Furthermore, these models are also useful in that they allow scientists to repeatedly perform controlled experiments on the computer to study the behavior of dynamical systems, such as the stock market, human or animal populations, or the earth's atmosphere, that would not be feasible to carry out in reality.

Beyond questions of utility, some scientists even claim that models such as cellular automata represent a complete paradigm shift in the way we conceptualize Nature, reflecting the universal principle that information underlies all interactions in the universe, and hence, all natural processes can be seen as computations. Though these nontraditional views have produced a lively and useful intellectual debate, it seems highly unlikely that theories such as cellular automata or complexity could ever replace Newtonian mechanics, for example, as a way of predicting planetary motion. As physicist Steven Weinberg has noted, despite being "interesting and important, they are not truly fundamental because they may or may not apply to a given system," and when they do, their underlying axioms cannot be derived from the truly fundamental laws of nature.[6]

4
Artificial Life

What is life? It is a flash of a firefly in the night.
It is the breath of a buffalo in the wintertime. It is the little shadow which runs
across the grass and loses itself in the sunset.
—Crowfoot, last words, 1890

If one were to take the themes that have arisen in the previous discussions of artificial neural nets and cellular automata to their logical conclusions, one might ask: could the fusing of biology and computer science be taken to such lengths that a computer could actually generate its own lifeform? Such beings would not merely be the simulation of a living creature, but rather be endowed with life in their own right within the internal ecosystem of the computer. These digital lifeforms could also in theory possess their own intrinsic biological principles that are just as valid, internally consistent, and capable of supporting the reproduction and growth of their species as carbon-based lifeforms on earth.

Life in Silico

Indeed, some mathematicians, information theorists, and biologists have made the claim that in principle, there is no real difference between life in vitro and life in silico. They assert that it's not the material composition of an organism that determines whether it's alive or inert (e.g., living tissue versus silicon), but rather the way it processes information.

Artificial Life, or "Alife" as its practitioners call it, is a field that explores the phenomenon of natural life by attempting artificially to create biological phenomena from scratch within computers and other nonliving media. A fundamental tenet is that Alife provides a more universal set of principles underlying the phenomena of life than those derived solely from traditional "earthly" biology, and makes them accessible to new kinds of manipulation and testing. The computer generation of life, claim Alife advocates, enables one to separate out fundamental principles from their earth-bound manifestations. In a sense, they say, the field represents the flip side of the empirical/analytic approach of most traditional biologists, who begin their study of life in a wet lab with a laboratory specimen—for example, a sea slug, mouse, or the common fruit fly, *drosophilia*—as their main point of departure. They then deconstruct it, analytically speaking, into its component parts (e.g., organs, tissues, cells, genes, molecules, etc.) in order to study its composition and behavior under different conditions, and over time, attempt, by an inductive process, to arrive at the fundamental organizational principles embodied there. If it were possible, of course, most biologists would prefer to simply reverse engineer life in the lab in order to grasp its principles, so they could themselves synthesize real organisms, cookbook-style, from their basic ingredients—but so far, this feat has proved elusive.

Non-Carbon-Based Life?

Biology is the scientific exploration of life here on earth. Despite its extraordinary complexity and diversity, the phenomenon of life was essentially the result of a historical accident. The first living organisms originated from a series of physicochemical reactions and environmental conditions that just happened to be prevalent on our planet at the time, in a completely causal chain of events. As it happened, the evolution of life as we know it today took a path based on carbon-chain chemistry.

There are some scientists who question whether a carbon base is a sine qua non for life. Could there be silicon-based life, or germanium-based life on a planet in another galaxy, as many science fiction writers have imagined? Or what if, at that very moment in history, over 3.5 billion years ago when the first life forms emerged, we had been able to slightly tweak the initial conditions that preceded them, and watch evolution begin again, only this time slightly altered? Would today's life forms be at all different? These are the sorts of provocative questions that can't easily be addressed in the laboratory, and that Alife purports to address.

According to Alife researchers—an eclectic mix of chemists, physicists, biologists, information theorists, and computer scientists—the study of carbon-based life as it evolved on earth constitutes a particular, not general, case, giving the necessary but not sufficient information for formulating life's basic principles shared by all living systems on all planets. In order to uncover these universal principles, one must explore the space of all possible biologies. But until aliens—whether iron-based or silicon-based—show up here on earth, most Alife practitioners feel the digital simulation of life can bring us closer to that goal. Santa Fe Institute's John Casti says, "I think Alife will ultimately enable us to properly understand evolution and the workings of cellular machinery mostly because it will offer us the chance to do the kinds of experiments that the scientific method says we must do—but cannot with the time and/or spatial scales of material structures like cells themselves."[1]

Alife scientists liken the artificial synthesis of life to the synthesis of chemicals in the laboratory, an endeavor that has enabled chemists to uncover some of the field's basic principles. A century ago, chemists' understanding of how matter behaved was derived primarily from what they could glean from the analysis of chemical reactions of substances they found around them. However, it would have been virtually impossible for them to arrive at an all-encompassing theory of chemical composition and behavior based solely on their observations of the behavior of a handful of chemicals chosen at random.

By the same token, those engaged in Alife believe that it is not until we understand how the universal concept of life could have appeared under many different circumstances and are able to actually synthesize life in a wet lab, will we reach this level of understanding. "To have a theory of the actual, it is necessary to understand the possible," remarked Chris Langton, one of the founders of Alife in the introduction to his first book on the subject.[2]

Alife proponents also cite the fact that the synthesis of new chemicals not found in nature has not only enabled scientists to broaden their empirical basis for study, as well as their understanding of underlying theory, but has produced socially useful products such as plastic, rubber, and pharmaceuticals. The promise of Alife is to be able to do the same with the life sciences, by harnessing some of biology's intrinsic principles in order to synthesize useful biomimetic products for industry and engineering—mirroring the extraordinary feats of bioengineering that nature has evolved on its own.

Some Alife scientists, such as biologist Charles Taylor at the University of California at Los Angeles, and computer scientist David Jefferson of Lawrence Livermore National Laboratory, feel that Alife can also help shed light on other aspects of biology. Alife modeling tools, in their view, are powerful enough to capture much of the complexity of living biological systems, yet in a form that is more tractable, manipulable, repeatable, and easier to subject to precisely controlled experimental conditions than their counterpart living systems. Other questions in science, according to Taylor and Jefferson, that Alife models could further elucidate, for example, are how cellular organization began; how ideas, terminology, or other entities of human culture evolved, while continuing to preserve their identities and to interact with one another; or how sex or gender has been maintained in organisms across the ages—a phenomenon that may involve interactions among genotype, fitness, parasitism, and so on. Alife could also aid our understanding of natural selection in large populations of organisms within their adaptive environments, claim Taylor and Jefferson, and in that context, how the concept of *adaptedness* relates to the *fitness* of the individual. It could provide new concepts and models of system phenomena emerging from complex or decentralized systems, such as flocking birds or the collective behavior of wasps as they build their nests. Other Alife proponents even express hopes that Alife will enable us to grasp the overall principles governing complex nonlinear systems.

Origins of Alife

Although many scientists have contributed to the development of Alife, the person who probably did the most to bring theoretical coherence to the field is Chris Langton. A brilliant yet iconoclastic scientist, partly an autodidact, Langton was preoccupied with creating the basis for artificial biological systems. His investigations finally took him to the University of Michigan in 1982, where he became a graduate student studying under Arthur Burks, the editor of von Neumann's papers on cellular automata, and John Holland, the father of genetic algorithms. This led to postdoctoral studies at the Center for Nonlinear Studies at Los Alamos National Laboratory in New Mexico.

Langton reached a point in his research in synthetic biology where he wanted to compare his ideas with those of others interested in information theory and biology, so he convened what became the first workshop on Alife. Sponsored by Los Alamos, the Santa Fe Institute, and Apple Computers, in September 1987, the gathering was called the "Interdisciplinary Workshop on the Synthesis and Simulation of Artificial Life." From the onset, the meeting's main purpose—to explore the concept of artificial life—was controversial. Even the program announcement read like a radical manifesto:

Artificial life is the study of artificial systems that exhibit behavior characteristic of natural living systems. It is the quest to explain life in any of its possible manifestations, without restriction to the particular examples that have evolved on earth. This includes biological and chemical experiments, computer simulations and purely theoretical endeavors. Processes occurring on molecular, social or evolutionary scales are subject to investigation. The ultimate goal is to extract the logical form of living systems. Microelectronics technology and genetic engineering will soon give us the capability to create new life forms in silico as well as in vitro. This capacity will present humanity with the most far-reaching technical, theoretical and ethical challenges it has ever confronted. The time seems appropriate for a gathering of those involved in attempts to simulate or synthesize aspects of living systems."[3]

The conference helped Langton's thoughts to coalesce. He was able to clearly articulate a description of Alife as the study of the "property of the organization of matter, rather than a property of the matter, which is so organized." Hence, the physical form of the organism wasn't as important as its behavior, such as self-reproduction, metabolism, growth, interaction, and adaptation. Moreover, he also settled on a second basic tenet of Alife: it was to be a *bottom-up* approach, similar to cellular automata, where localized behavior could create globalized emergent behavior, as is characteristic of self-organizing systems.

Alife and the Individual Entity

Alife, as a field of scientific endeavor today, has been applied to problems in many disciplines, such as engineering, computer science, biology, physics, chemistry, sociology, and even economics. In his seminal textbook on subject, Caltech physicist Chris Adami places Alife research along a lengthy continuum with the simulation/emulation of individual entities at one end, and the emergent properties of large populations of individuals at the other. As an example of the former, he gives Karl Sims's simulation of the evolution of the form and movements of swimming behavior in virtual sea animals made out of simple blocks, also illustrating how competitive behavior eventually emerges in these creatures (figure 4.1), or the localized behavior of wasps when building a nest that produces a variety of symmetric, architecturally sophisticated structures, even though each wasp is only aware of its own building behavior.[4]

An analogous example in engineering would be the construction of adaptive, autonomous robots, whose software enables them to interact with their environment, and evolve and learn from it. This method, as most techniques classed as Alife, employs a *bottom-up* approach, where the robot's cognitive system is constructed from simple elemental units that gradually develop through evolution and adaptation to their environments into more complex systems. This contrasts to robots constructed with artificial intelligence (AI) that employ a *top-down* approach, where a complex behavior (such as walking up steps) is targeted from the onset and the system is designed and built already containing all the elements it needs to achieve the detailed, complex behavior. What's more, AI has traditionally focused on machines being able to perform complex, multifaceted human functions, such as chess playing, voice comprehension, or even medical diagnosis, whereas Alife looks exclusively at natural behaviors, emphasizing survivability, evolution, and reproduction of the creature in complex, dynamic environments.

An even more intricate example of engineered virtual life is the Golem (Genetically Organized Lifelike Electro Mechanics) project at Brandeis University, involving robots that can actually design and build other robots. Devised by computer scientists Hod Lipson and Jordan Pollack, the parent bots consist of a computer running an evolutionary algorithm that produces a design based on natural selection, and a 3D printer that makes small plastic shapes. The offspring are small plastic trusses with actuators, propelled by motors and controlled by artificial neural nets (figure 4.2). Humans intervene only to attach the motors and connect the wires—the robots do all the rest, including telling the humans what to do. More recently, these scientists have been expanding

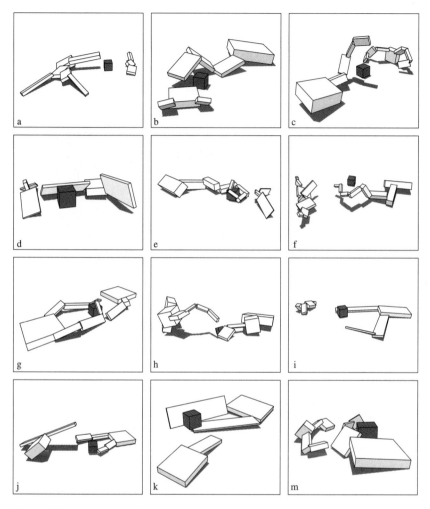

Figure 4.1
Karl Sims's creatures that have evolved to compete with each other. (Image courtesy of Karl Sims, 1994, with permission of author.)

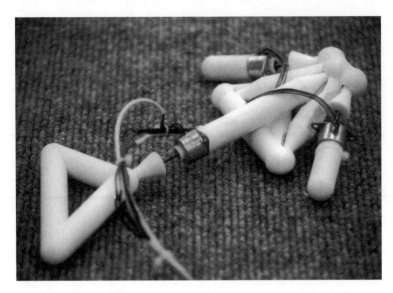

Figure 4.2
An image of a Golem project's computer-designed evolvable robot, which pushes itself along the carpet using the piston at the center. (Image courtesy of the Golem Project at Brandeis University.)

the evolutionary process so it can produce increasingly complex machines and also 3D–print the wires and batteries preassembled into the robots.

Alife and Emergent Behavior

At the other end of the Alife spectrum, Adami places the study of populations that display globalized properties that can't be seen in the behavior of the individual units, known as "emergent behavior." (For example, in physics, temperature and pressure are examples of emergent behavior—also called state variables because they describe the state of the system—ascribed to a large system of interacting molecules. An individual molecule has neither temperature nor pressure by itself.) According to Adami, because these living systems aren't amenable to a statistical description in terms of macroscopic variables, one might attempt to describe the behavior of the system's individual members in parallel to try and view its emergent behavior. This approach falls short, though, because parallelism is unable to capture the self-organization that appears only in the aggregate in many living systems, and also because the resulting emergent behavior, through a kind of feedback loop, can affect individuals in nonlinear ways, producing unexpected results.[5] Examples of this are swarm intelligence that emerges when termites build complex systems of tunnels in wood, without each insect really knowing what it's building. Adami also cites the example of Craig Reynolds's work on flocking birds, which studies how flocks of birds fly without instructions from a leader. Reynolds created a virtual flock of birds, called "boids," which flew according to three rules:

- Always avoid collisions with your neighbors;
- Always try to fly at the same speed as your neighbors;
- Always try to stay close to your neighbors.

These three rules were sufficient to create the emergence of flocking behavior. The boids flew as a coherent group, automatically spliting into two groups when encountering an obstacle, and reuniting after it had passed (figure 4.3). This demonstrates how a system of fairly simple

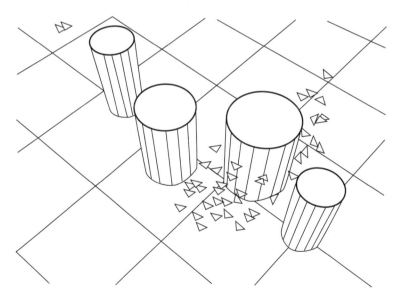

Figure 4.3
Craig Reynolds' "boids" automatically split into two groups when encountering an obstacle, and reunite into a coherent group after passing through it. (Image courtesy of Chris Adami, Digital Life Laboratory, Caltech.)

elements interacting with their nearest neighbors, with no central direction, can create cohesive, intelligent group behavior. Indeed, the behavior of Reynolds's boids mirrors what one actually observes in nature.

Life on Tierra

Somewhere between these two extremes lies Tom Ray's Tierra project, which explores open-ended evolution within a virtual world, unfolding without any a priori human instructions. Ray modeled Tierra on the period in Earth's evolution known as the Cambrian Era, about 600 million years ago. This era began with the existence of simple, self-replicating biological organisms that underwent explosive growth over time to result in the great diversity of species known today. Ray wanted to investigate how self-replication eventually produced such a wide variety of complex lifeforms. He started with a single organism called the "Ancestor," the only engineered lifeform in Tierra. Then he unleashed his virtual creature and watched to see what would happen. After only one night, his virtual world was teeming with myriad creatures, displaying an amazing variety of form and behavior. These organisms and their progeny (the "organisms" are actually self-reproducing programs written in assembly language) competed for the natural resources of their world, CPU time and memory. This provided the basis for natural selection to operate—some organisms die off, and the fitter ones survive, adapt, and become more competitive. In order to prevent these digital beings from gaining access to the actual hardware of the machine they lived in, like computer viruses, Ray made the entire Tierra program run on a virtual computer created in the software. The operating system of Tierra performed four functions: (1) allocation of memory to each organism, in addition to the exclusive privilege of modifying its own structure, in order for it to preserve its unique identity; (2) allocation of CPU time to each organism in order for it to act; (3) placement of each organism in a queue, and, depending on life cycle, natural selection, and so on, killing it off when it reaches the top of the queue; and (4) doling out random mutations in the binary string of each organism's program, thus causing some organisms to self-replicate imperfectly.[6]

Ray, currently a zoology professor at the University of Oklahoma, made his first version of Tierra in January 1990. In its first run, after over 500 million instructions in the program, the Tierra program had evolved more than 350 different lifeforms, ninety-three of which were fit enough to survive to achieve subpopulations of five or more individuals. Tierra also generated hosts and parasites, and eventually a type of social organization with communities of genetically uniform organisms emerged. Nearly every facet of natural evolution and every form of behavior attributed to lifeforms here in earth appeared in Tierra, including competitive, exploitative, and protective behaviors.

Life among the Avidians

Inspired by Ray's work, Caltech's Adami and Richard Lenski, a microbiologist at Michigan State University, began their own Alife experiments a few years later. Lenski had been conducting (and continues to conduct) wet lab experiments on evolution in his lab with *E. coli* bacteria, where a single experiment can span up to 24,000 generations, usually producing a new generation about every 3.5 hours. Lenski happened to read Adami's book on Alife, and soon after contacted him to learn more about the field, including a new software program called "Avida" that Adami's students Titus Brown and Charles Ofria had developed. Avida was created not only to enable digital lifeforms to evolve, as they had in Tierra, but to subject them to various experiments in the course of their virtual evolution, such as inserting a particular type of mutation and seeing what resulted in the ensuing generations. The Avida program typically ran for about 5,000–20,000 generations, taking only from two to eight hours to complete, compared with similar experiments that could take days performed in vivo or in vitro in a lab.

Adami and Lenski then collaborated to design a virtual environment where one initial virtual bug was programmed to be able to reproduce itself, mutating every thousand or so "births." The bug could also perform certain simple mathematical functions that the environment rewarded by allowing the successful bugs to replicate at a faster rate, thus pushing out the less competent ones and forcing them eventually to die off—just as natural selection does in a real environment. According to Adami, he and Lenski subsequently carried out experiments with both the virtual Avidians and *E. coli* in vitro, to be able to compare the results. They have found not only that many of the evolutionary principles seen in the computerized environment quite accurately mirror those found in nature, but also that the survivors in their system appear stronger and less affected by random mutations than the less fit individuals. The experiment ostensibly puts virtual and actual results on the same footing, in order to test how accurately Avida can replicate the workings of natural evolution.

This comparison, at first blush, may seem like an effective way to validate some of the claims of Alife; however, University of Colorado evolutionary biologist Rob Knight believes the issue is more subtle:

One must consider that a major distinction in the Alife community is between "weak" and "strong" Alife. Weak Alife, which claims that simulations of evolving systems may help us understand *real* biological life, is relatively uncontroversial, especially when the simulations closely mirror natural systems, such as Nilsson and Pelger's study of the evolution of eye morphology.[7] These models clearly provide valuable insight. Weak Alife proponents, however, are loath to ascribe more to these simulations than being simply a modeling tool. Even more skeptical are the majority of experimental biologists, who tend to distrust computer experiments, perhaps unfairly, on the grounds that models often merely reflect biases programmed in at the start. Then there are strong Alife proponents, who claim that replicating programs inside computers really are alive in their own right. Their views are far more controversial. This is partly because the examples of biological life we are all familiar with are orders of magnitude more complex, and partly because the claimed similarities with biological evolution tend to be rather abstract.[8]

From these comments, it would appear than Lenski and Adami belong to the latter group, as they give equal weight to both their digital and in vitro life forms.

Alife versus Life Itself

The question of the nature and origin of life are weighty ones, having preoccupied human thought since the dawn of time. Priests, shamans, philosophers, poets, and naturalists have all addressed it; all the theologies and all the mythologies of the world have given it a central place in their belief systems, but only lately in the history of mankind has it also become the purview of science, creationism notwithstanding. Given the import and ramifications of this question, it's not surprising that the idea of computers creating their own bonafide life forms should elicit debate, skepticism, and even incredulity. Alife represents a new multidisciplinary field that appears to be thriving and attracting new followers. When judged on its own merits as a fascinating new subset of computer science rather than as the study of meta biology, Alife has already produced a significant body of research, and its scientists are having an impact in a variety of areas, such as algorithm development, software, hardware, robotics, population biology, economics, and complex systems analysis.

To go the next step and view Alife creatures as a completely authentic and *real* form of life, possibly revealing a universal theory of biology that transcends anything seen on earth, may, however, require a leap of faith. Some critics and skeptics involved in a kind of cultural war with theories of this type, tend to dismiss Alife as a faddish trend that few mainstream biologists worth their salt would ever take seriously as anything other than a modeling tool. When viewed as a simulacrum for real biology, most traditional life scientists, particularly evolutionary biologists, feel Alife could never be a legitimate vehicle to study living organisms or how they evolved. When asked for his views on Alife, chemist Stanley Miller, who in 1952 conducted the first experiment that simulated the primordial soup of the primitive earth and produced amino acids, purportedly declared, "Running equations through a computer does not constitute an experiment!" However persuasive these arguments, judging the potential or merits of Alife solely on the basis of how well its results compare with those derived from traditional biology may be using unfair criteria and diminish its overall accomplishments.

Biochemist and engineer Andy Ellington from the University of Texas at Austin, who studies evolution in his lab, believes the nub of the problem may lie with confusion over how we define life. Says Ellington:

The whole notion of *life* is somewhat specious. It is frequently difficult to draw meaningful scientific distinctions between organisms, viruses, and growing crystals. Thus, I have no problem with those who say that life inside a computer is "real" life; the word is as ambiguous inside a computer as outside. From this standpoint, while creatures (or replicators) spawned by Alife couldn't adapt to a biological environment (imagine a cellular automata in a rain forest), they can compete and adapt in a virtual one. So artificial life shouldn't be judged on the basis of whether or not it's as valid as biological life, but should be regarded as a completely separate entity, which a priori doesn't need to have the same underlying rules as biological replicators. What we may need is a paradigm shift in the way we think about the concept of life.[9]

Rob Knight agrees, and believes that in order to convince the skeptics, Alife researchers need to define more clearly their view of *life* and its characteristics, and carefully demonstrate how Alife experiments can both recapture and extend the results of more traditional methods of inference about evolution.

Alife has attracted a multidisciplinary following of talented researchers bringing vitality and originality to their work. The field has spawned some extremely novel varieties of in silico life and provided a testbed for ideas that defy laboratory experimentation, in addition to spinning off novel software and other modeling tools. Alife's controversial nature has caused proponents and critics alike to re-think some of their often unexamined assumptions about the nature of life, ultimately producing a rich and provocative corpus, which—whatever one's judgement—has the potential for continuing to generate new ideas and directions for research.

5

DNA Computation

Transparent forms, too fine for mortal sight, Their fluid bodies half dissolved in light.
—Alexander Pope, "The Rape of the Lock"

Our genetic material, or DNA, encodes information in the form of a particular sequence of molecules, and is often compared to computer software. The information it contains, which plays a major role in determining our hereditary traits, gets "processed," so to speak, through biochemical means in order to control the synthesis of proteins within the cell. The fact that genes are a means for encoding and processing information is at the crux of the idea behind DNA computation.

DNA is made up of four different kinds of molecules called nucleotides, which are organic compounds composed of a nitrogen-containing base, a phosphate group, and a sugar molecule. Their particular sequence or way they are ordered represents a code that spells out the biochemical instructions for the production of a given protein, just as a computer software program encodes instructions based on a pattern of zeros and ones.

The mode in which a segment of the DNA code gets activated to produce a specific protein depends on any one of a series of complex biochemical operations that serve to splice, copy, insert, or delete the segment. These operations are analogous to the arithmetical and logical operations performed on the zeros and ones contained in a computer program.

DNA computation makes use of this ability to store and manipulate the information in a genetic code to devise algorithms for solving a problem, in some ways even improving on what a conventional computer could do for a certain small group of problems, claim some. Not even a decade old, DNA computation has grown considerably as an interdisciplinary field of research since the first conference in 1995, and has given rise to some interesting new avenues of research such as the self-assembly of DNA molecules, which could prove a useful tool for building structures for nanotechnology. DNA also has appeal as a storage medium—for example, a cubic centimeter of DNA stores about 10^{21} bits of information. It can perform trillions of operations in parallel—far more than any electronic computer—and uses about 10^{-9} of the energy of conventional computers.

However, DNA computing is not without its limitations, and will likely never function for general purpose computations (in other words, it's doubtful that in thirty years' time, we'll be using vials of DNA on our desktops to compute the solutions to problems). Although, as a method, it has been successful at solving small-scale search problems using combinatorial algorithms, it has been unable to scale up for larger problems, as such exponentially large amounts of DNA would be needed that the computational techniques wouldn't hold up. In addition, DNA computing even for medium-scale problems requires very painstaking laboratory work to implement, and as a consequence, usually entails a significant amount of error. Whether or not it will find its "killer ap"— the application for which it's the best means possible, transforming DNA computation into the world's next revolutionary computing device or a promising new information storage medium—is still uncertain. Some critics have also questioned the method's practicality because of the highly laborious lab work it entails and the high cost of some of the materials required.

That said, DNA computing has broken completely new ground in science—an rare achievement in itself—in addition to having spun off new research areas such as DNA self-assembly, and attracted many talented researchers to the field. Its cross-disciplinarity has lured computer scientists into the biology lab, where they have had to master the techniques of recombinant DNA, while providing their biologist colleagues with a synergistic perspective on the computational properties of living cells, introducing them to concepts such as networking, systems, input and output, switch, and oscillator that can be applied to cellular functions.

Genes, Nucleotides, and Computing

The work of mathematician Tom Head at SUNY Binghamton in the late 1980s served as an important precursor for the discovery of DNA computing. Head devised a way to use DNA sequences as a code for composing and executing algorithms. He developed a *splicing model*, experimenting with recombinant DNA lab techniques on DNA strands to carry out simple rules taken from formal language theory.

The actual origin of DNA computing can be dated in time to the summer of 1993. Its discovery so stirred the imagination that accounts of it appeared in leading newspapers and science magazines the world over. The idea for computing with DNA was the inspiration of theoretical computer scientist and mathematician Leonard Adleman of the University of Southern California as he was perusing a biology textbook. Adleman already had a highly distinguished scientific track record as co-discoverer of a way to implement public-key cryptography in the late 1970s, while a young faculty member at MIT. (Cryptography is a way to encode messages to make them unintelligible to anyone other than those authorized.) His work, a collaboration with MIT colleagues Ron Rivest and Adi Shamir, became the basis for RSA encryption, now a widely used international standard for encrypting information, especially over the Internet. It also propelled the three scientists to worldwide fame, and led them to create a successful high-tech company, RSA Inc., to handle all business matters related to RSA.

In the 1980s, Adleman had begun to study biology and learn the methods of wet lab experimentation, being attracted to the field by the unsolved problem of HIV/AIDS. It was while learning about the biomolecular mechanisms underlying in this condition that he read a section in a textbook about polymerase, the enzyme that catalyzes the elongation of a new DNA strand during DNA replication. He was struck by the similarities between the way both DNA and computers process encoded information, and it occurred to him that, by cutting, pasting, and copying DNA in the right order, it should be possible to carry out certain simple algorithms with strands of DNA. He then proceeded to astound the scientific community by actually carrying out a difficult computation, called "The Directed Hamiltonian Path Problem," on pieces of DNA in test tubes, using the standard laboratory techniques of recombinant DNA or genetic engineering. His accomplishment generated widespread intellectual excitement among scientists because of its revolutionary nature and the seeming potential of DNA to outperform conventional computers. Some scientists saw its power in its ability to encode information in such a small space, and by manipulating the strands, to sort through an exhaustive library of all possible answers to problems of a certain size.

The Gene-Based Computer

DNA is a major constituent of the chromosomes that determine our heredity. It consists of two long molecular chains twisted into the form of a double helix, and joined by hydrogen bonds (figure 5.1). The bonds are linked to four nucleotides, adenine and thymine (A and T) or cytosine and guanine (C or G). In living organisms, the bases in each of these nucleotides bonds with its complement—A to T and C to G—in a pattern that ultimately determines our heredity. (A base is a chemical compound, distinguished from an acid, that contains oxygen and hydrogen bound together to form a hydroxyl ion, and has a pH greater than 7; a common base is sodium hydroxide, NaOH.) Two DNA strings are complementary if the second has a sequence such that A and T are interchanged, and C and G are interchanged.

In order to compute using strands of DNA, one must first start by elucidating, step by step, the computer algorithm that solves the problem of interest. Then, one must translate each of these steps into the language of DNA, by determining the sequences and exact manipulations to be performed on the DNA. The DNA starting material with the proper sequences must be custom made in a lab; these short strands are known as oligonucleotides. This process usually involves the close collaboration of computer scientists and biologists; the cross-pollination that typically results has helped foster growth in the field.

The DNA-based computation can be arranged as a series of test tubes filled with water and up to 10^{20} strands of DNA. The strands consist of the preselected sequence of oligonucleotides, which are then manipulated and sorted according to the algorithm. For example, they can be sorted by length, or by performing various "and," "or," or "not" operations to check for the presence of certain sequences in an individual strand of

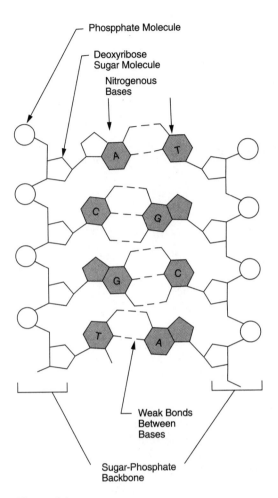

Figure 5.1
DNA structure. The four base pairs of DNA are arranged along the sugar-phosphate backbone in a particular order, encoding all genetic instructions for an organism. The bases adenine (A) pairs with thymine (T), while cytosine (C) pairs with guanine (G). The two DNA strands are held together by weak bonds between the bases. (Source: US Department of Energy Human Genome Project, <http://www.ornl.gov/hgmis>)

DNA. The strands remaining at the end of these operations are amplified using standard molecular biology lab techniques, and then are sequenced to identify the resulting DNA sequence that represents the answer to the problem (figure 5.2). In principle, only one strand containing the answer needs to be amplified.

Each operation performed sequentially on the DNA creates a new batch from the results of earlier mixtures; the operations can consist of separating the strands by length, pouring the contents of one test tube into another, extracting those strands with a given sequence, heating or cooling them, or using enzymes to splice the DNA. The series of test tube manipulations forms what computer scientists call "a single-instruction, multiple-data" (SIMD) computation performed in parallel on the DNA.

The numerous laboratory techniques of molecular biology, known as recombinant DNA techniques, enable a broad range of algorithmic operations to be used in DNA computing. These methods have been developed over the last few decades for use in biology wet labs for genetic engineering purposes. The principle laboratory operations used in DNA computing are the following:

• Synthesizing DNA strands of a desired sequence of A, T, G, or C and length.
• Separating double-stranded DNA into single strands by heating, breaking the hydrogen bonds that connect them. This is called *denaturation*.
• Fusing single-stranded DNA into double-stranded DNA with complementary base pairs connected—called *annealing or hybridization*.
• Removing all the single strands or all the double strands from the mixture.
• Cutting the strands in exact places using restriction enzymes. (An enzyme is a catalyst that causes a chemical reaction to occur.) Many bacteria produce restriction enzymes, protecting the cell by cleaving and destroying the DNA of invading viruses. Each restriction enzyme cleaves the DNA at a particular site, so each operation requires a different one.
• Separating the strands by length.
• Extracting and separating out strands with a known sequence of 15–20 base pairs anywhere along the strand.
• Finding a particular string encoded in a DNA strand.
• Constructing complementary strands for short (15–20 base pairs) strings in a strand.

1. Sequencing reactions loaded onto polyacrylamide gel for fragment separation

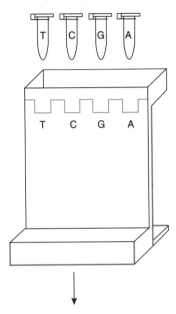

2. Sequence read (bottom to top) from gel autoradiogram

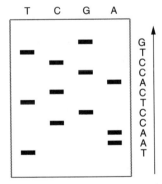

Figure 5.2
DNA sequencing. This particular method of sequencing (called the Sanger method) uses enzymes to synthesize DNA chains of varying lengths, stopping DNA replication at one of the four bases and then determining the resulting fragment lengths. The fragments are then separated by a method called electrophoresis (1) and the positions of the nucleotides analyzed to determine sequence (2). (Source: US Department of Energy Human Genome Project, <http://www.ornl.gov/hgmis>)

• Making many copies of a given strand of DNA using the polymerase chain reaction (PCR). To use PCR, double-stranded DNA is heated and separated, and then short strands that are complementary to sequences in the targeted DNA strand (primers) are added, together with free base pairs and the enzyme DNA polymerase, which catalyzes the elongation of a piece of DNA. In a series of heating and cooling cycles, the DNA doubles with each cycle.

• Appending a given string of DNA to a selected substring, or to the entire strand of DNA present, called *ligation*.

• Destroying a marked strand of DNA.

• Detecting and reading. This enables one to determine if a test tube of DNA does/doesn't contain at least one strand with the desired result; and, if it does, to interpret its sequence.

The Traveling Salesman

Adleman's first successful test of the theory of DNA computing was solving a well-known problem in computing called the Directed Hamiltonian Path Problem (or "Traveling Salesman Problem"). According to Adleman's method, a particular DNA sequence can represent the vertices or endpoints of a graph, and one can describe random paths through the graph by performing operations on the various strands of DNA. In this case, Adleman chose seven cities to represent the vertices of the graph, and fourteen links connecting the cities in various ways that represent one-way non-stop flights between two cities. The problem entails figuring out whether there is a route that takes the salesman from a given starting city to a given end city, passing through each city once and only once (figure 5.3).

Step one is to assign each city a unique six-letter name, where each letter stands for a given nucleotide (adenine, guanine, cytosine, and thymine, or A, G, C, and T). So in this case we do the following:

San Francisco = AGCTAT
New York = GTTAGC
Seattle = ACACTA
Sydney = AGTTAT
Seville = TATCTC
London = TAAAAG
Rome = GATGCT

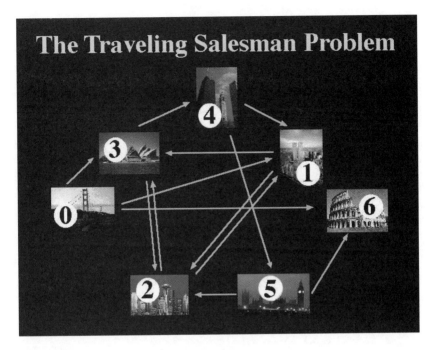

Figure 5.3
DNA computing solved the seven-city directed-path problem by reacting coded DNA strands, as represented by the letter sequences next to the pictures, to imprint a successful route on the product material. (Source: L. Kari and L.F. Landweber, "Computing with DNA," in *Bioinformatics Methods and Protocols*, S. Misener and S. Krawetz, eds., Humana, Totowa, NJ, 1999, p. 415)

Then one assigns the corresponding flight names for those cities which have direct connections, using the first three letters of the city name to signify arrival and the last to indicate departure, for example:

Departs San Francisco, arrives Sydney = TATAGT
Departs Sydney, arrives London = TATTAA

and so on. The four nucleotides only bond chemically with their complementary base pair, so each city code also possesses a complementary name, for example:

San Francisco (original code) = AGCTAT
San Francisco (complement code) = TCGATA
Sydney (original code) = AGTTAT
Sydney (complement code) = TCAATA

Single strands of DNA containing all the direct flight names and other strands containing all the complementary city names are then synthesized in a lab and mixed together in a test tube, so if a strand containing a flight from San Francisco to Sydney encounters another with the Sydney complement city name, the two will bond:

```
TATAGT
   | | |
   TCAATA
```

Then, if this strand bonds in the test tube with one containing a flight from Sydney to, say, Seville, we have:

```
TATAGTTATTAT
   | | | | | |
   TCAATA
```

and when this piece runs into another containing the Seville complement name, we get:

```
TATAGTTATTAT
   | | | | | | | | |
   TCAATAATAGAG
```

which will react with a strand for a direct connection between Seville and another city, and so on. The next step is to read out the strands containing the answer, which results from first filtering, measuring, and reading out the DNA in the test tubes so that only strands beginning with San Francisco and ending with Rome, and that contain all the five other city codes and are seven city codes long, remain in the test tube. The sequence in these strands will contain the answer.

The remarkable parallelism of the DNA computer uses brute force to try out all the possible solutions to the problem. It does so in about 10^{14} operations per second (assuming the binding of two DNA molecules is a single operation, and also that approximately 10^{14} copies of the associated oligonucleotide are added to represent each direct flight, and that half of these bind).

The traveling salesman problem forms part of a class of problems called "NP-complete," where "NP" stands for non-deterministic polynomial time. This means a set of problems whose answers can be checked in polynomial time, that is, where the computer running time needed to verify if an answer is correct or not is bounded by a polynomial function. (This is assuming that a genie has already revealed the correct answer to be checked!)

NP-complete problems, a special subset of NP problems, currently have no known polynomial time solution. What makes Adleman's accomplishment so remarkable is that by successfully demonstrating that the traveling salesman problem could be solved with DNA, he showed, as proof of concept, that the biological computer was capable of solving a small-scale NP-complete problem. (This does not mean, however, that all NP problems can be feasibly solved.)

Adleman's experiment set a completely new direction in bio-inspired computation. Even though his approach could not be scaled up indefinitely to very large combinatorial search problems as the number of individual DNA strands needed tends to grow exponentially with the size of the problem, it was a successful demonstration of a paradigm shift in computing. Moreover, it showed that individual molecules could be manipulated in the lab to implement complex algorithms—a result that those in the emerging field of nanoscience found very encouraging.

The Fledgling Field Advances

Adleman's experiment helped attract many talented scientists to the field from biology, computer science, physics, chemistry, and engineering, and growth continues in both theory and experiments with DNA computing. One type of problem that several researchers have found very amenable to DNA computations is the satisfiability problem (SAT), an important NP-complete problem in computer theory.

Solving SAT problems with DNA demonstrates how DNA computers work as powerful search engines. Satisfiability problems attempt to answer, for a given Boolean expression A, if values can be assigned to the variables in A that would make A a true statement. With DNA, any strand that encodes values that make the expression true is a potential solution. For an SAT problem with n variables, an electronic computer must test 2^n variables one by one, which can require a lot of time for large n. One theoretical approach to SAT with DNA proposed by Georgia Tech computer scientist Dick Lipton, however, can check all the variables simultaneously, providing a much more efficient method for solving the problem.

Lloyd Smith and colleagues at the University of Wisconsin have used a slightly different technique, called surface chemistry, for carrying out SAT problems with four variables, involving sixteen possible truth assignments. This method entails placing DNA that encodes all the possible Boolean variables onto surfaces, using restriction enzymes to destroy those strands that don't satisfy the Boolean formula, and reading out the results with an optical fluorescence technique.

Biochemist Kensaku Sakamoto of the University of Tokyo solved a six-variable SAT problem using "hairpin DNA," which takes advantage of DNA's tendency to get tangled and tied up in knots. Instead of adding various enzymes to destroy the DNA strands that contain the wrong answer, Sakamoto and his group designed their experiment so that, when cooled, the strands with the wrong answers would automatically fold over and form a DNA hairpin. The method, though unique, is still being refined to reduce the number of errors in the final answer.

Princeton molecular biologist Laura Landweber and colleagues solved a nine-variable SAT problem using a combination of DNA and RNA techniques. This problem is related to the "Knight's Problem" in chess and involves a possible 512 truth assignments. Len Adleman and his colleagues recently solved a twenty-variable SAT problem in the lab, after using DNA computation to exhaustively search more than a million possible answers. Performing this computation with twenty variables, the highest number reached so far, involved overcoming some difficult problems with errors. Those in the field predict that solving eighty-variable SAT problems with DNA may be the ultimate limit because at that point there is exponential growth of the volume of liquid containing the DNA. All SAT experiments require preparing a DNA "word library" beforehand, containing strands encoded with all possible truth assignments.

Another significant advance came in 1997, when Lipton and two of his graduate students devised a theoretical method for using DNA to break the data encryption standard (DES), developed by the National Security Agency. Data encryption standard makes use of a single key, among 2^{56} keys, to scramble messages. In order to break the code, one must test each of the 2^{56} keys one by one, which would take an *extremely* long time on a conventional computer. However, in the Lipton et al. approach, all the possible keys can be encoded in strands of DNA and, through a series of steps involving extractions, replications, and other biological processing operations, they can all be searched simultaneously for the correct one. Breaking the first key would take months; however, subsequent keys can be cracked in a matter of minutes. Although Lipton's method may never actually be implemented in the lab, it does underline DNA's massive parallelism and its capacity for information storage (DNA-based computers use about a trillionth of the space that conventional computers do to store data).

Some researchers, like Duke's John Reif, have proposed using DNA as a voluminous database for information storage; it makes use of a recombinant DNA technique called the polymerase chain reaction (PCR) as a search engine. Polymerase chain reaction reproduces a fragment of DNA so as to produce many copies of the fragment. (The technique has proved an invaluable method in biotechnology, especially in forensic science, enabling amplification of minute traces of DNA for DNA fingerprinting.) Scientists at India's Central Scientific Instruments Organization have recently developed another method for DNA data storage, involving an innovative software program that encodes digital information into DNA sequences, not only for storage but also for computation.

Many of the field's proponents hope that a successful killer ap for DNA computing will ultimately be discovered, to spur the field on to the next important stage in its evolution, with additional research funding and even commercial support.

Desktop DNA?

Like any fledgling technology, especially one as novel as DNA computing, the field has its skeptics who contend that, as a new method of computation, it has claimed more than it can deliver. However, if judgments be made, a DNA-based computer should not be assessed for its usefulness as a general-purpose computer, or as a future replacement for the silicon-based PC, but only for very specialized applications. So far, DNA computing seems best adapted to small-scale search problems, and to problems that exploit its vast potential for data storage.

Other scientists—perhaps in response to some of the media hype that surrounded the field when it was new—cite the quantity of errors in the answers that stem from the difficulties in handling a DNA computer. For instance, to read out the final result, one may want to extract only the DNA with a specified pattern from the test tube, and error rates for extraction typically are about one part in a million. Another source of error results from the decay of DNA, which particularly affects computations that take several months to do. However, reseachers are currently working to demonstrate that these error-prone computations can be re-arranged so that they are virtually error free.

DNA Computing in Vivo: The Ciliate

The DNA computing described so far takes place in the lab, and can be considered an in vitro (literally, in the glassware) operation. Computing inside tiny organisms known as ciliates provides a good example of in vivo computation. Ciliates are single-celled organisms (ciliated proto-zoans of genus *Oxytricha* or *Stylonychia*) that live in pond water, and swim and feed with the aid of a layer of fine hairs, or cilia. They have been on earth for about two billion years. Ciliates have two different types of nuclei, the structure inside the cell containing the genetic material that determines all the cell's structures and functions. In the smaller nucleus, all the genetic material is jumbled, fragmented, and broken into smaller segments, and includes what is known as "junk DNA," which apparently serves no purpose. Scientists have discovered that some of these genes must be unscrambled in order to form the second, larger nucleus, where the genetic instructions are reorganized so that they carry out their functions properly.

Princeton molecular biologist Laura Landweber, theoretical computer scientist Grzezorg Rozenberg at the Leiden Center for Natural Comput-ing at Leiden University in the Netherlands, and ciliate expert David Prescott at the University of Boulder, Colorado, have studied what it takes for a ciliate to reassemble a scrambled gene, how the programs are written, and how much backtracking and error-correcting the process involves. Rozenberg calls this new computational paradigm "computing by folding and recombination." He has used elaborate constructs to help discern the logical rules followed by the ciliate in untangling its genes. Understanding this process might help scientists grasp how random mol-ecules in the primordial soup eventually organized themselves into the genetic code.

Whither DNA Computers?

In the final analysis, there's always a risk in trying to predict how any particular scientific development will fare in the future and what its impact will be, particularly something as unique as DNA computing. However, according to originator Len Adleman, its ultimate success or failure even as a special purpose computer may even not matter. He observes that

whether or not DNA computers will ever become stand alone competitors for electronic computers—which is unlikely—is not the point. Every living cell is filled with thousands of incredibly small, amazingly precise instruments in the form of molecules, which comprise "Nature's Tool Chest," a tool chest for the 21st century. These molecules store information, store energy, act like motors, or like structural material, and can cut and paste. Each is incredibly small, extraordinary precise, and functions with an energy efficiency that is on the cusp of what is thermodynamically feasible. I believe things like DNA computing, along with the other ways we are learning to use these wonderful tools inside the cell, will eventually lead the way to a "molecular revolution," which ultimately will have a very dramatic effect on the world.[1]

6

Biomolecular Self-Assembly

It has long been an axiom of mine that little things are infinitely the most important.
—Sir Arthur Conan Doyle, "A Case of Identity in the Adventures of Sherlock Holmes," 1892.

Scientific instrumentation and experimental methods in the life sciences, materials science, physics, chemistry and other scientific disciplines have advanced radically in the last forty years, continually improving our ability to observe, handle, and measure matter at smaller and smaller dimensions. At present, tools like the scanning tunneling microscope can take us down to the scale of nanometers, or roughly 1–100 billionths of a meter (the diameter of a human hair is about 73,000 nanometers), where one can actually "see" the atoms and molecules that make up matter. These developments have opened up a new domain for science and technology, called nanotechnology, which involves the manipulation of matter at atomic and molecular levels. The field is perforce multidisciplinary because at these dimensions, classical and quantum physics, chemistry, materials science, biology and engineering all come into play.

Nanotechnology does not simply mean mechanical engineering that has descended one or two more rungs down in dimensionality, where the components are smaller but still behave as we know materials do in the everyday world measured in meters. On the contrary, because we are now reaching atomic dimensions, the properties of matter at this scale can often seem counterintuitive and fly in the face of the observed behavior of materials at the macroscale. At the nanoscale, the quantum effects of matter begin to dominate. For example, if a magnet is placed on a piece of superconducting metal at room temperature, and the metal is slowly cooled down to the superconducting temperature, the magnet will begin to levitate, as if suspended in thin air. This effect, called the Meissner effect, is due to quantum mechanical properties of superconductors that effect the magnetic fields in and around them (figure 6.1).

Consequently, as the tried and true methods for fabrication, such as cutting, welding, or machining, are completely inadequate at this scale, we can look to nature to inspire new ones. In fact, we may not have to look any farther than inside the living cell, which contains many of its own nanomachines performing various biological functions.

This new science of working with matter at molecular dimensions has implications for applied research and technology in areas such as computer logic, biomedicine, "designer" materials created directly from atoms or molecules, and new nanoscale instruments that can probe the interior of cells without disturbing their behavior. Through the manipulation of matter, atom-by-atom, as if it were LEGO blocks, and the accumulation of skills and knowledge learned in the process, we will eventually be able to scale up these modular nanostructures into systems, whose design, control, and construction will ultimately remain at the nanoscale. In addition to leveraging what we learn about naturally occurring intracellular "machines," contact with new phenomena in chemistry, material science, and physics will also contribute to the learning process, leading to new insights as well.

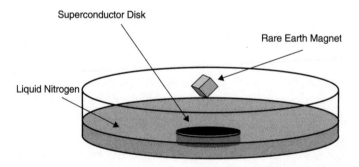

Figure 6.1
The Meissner effect. As a piece of superconducting metal gets cooled down to superconducting temperatures using liquid nitrogen, it has an effect on the magnetic fields around the metal, causing a magnet resting on its surface to levitate.

Self-Assembled Nanostructures

An experimental technique known as molecular self-assembly is proving to be an essential tool for the design and construction of nanostructures. *Self-assembly* is the autonomous formation of natural—that is, not man-made—components into organized structures.[1] The basic building blocks for these structures are atoms, molecules, or groups of molecules. Although self-assembly techniques, strictly speaking, tend to be associated with molecular structures at the nanoscale, insofar as it describes the collective behavior of any group of similar objects, the term can also be applied on the macroscale to crystal growth, small-particle emulsions, snowflakes, schools of fish, flocks of birds, or even meteorological phenomena such as cyclones.

In self-assembled objects or phenomena, the component parts must be preformed, and the assembly process must take place without any human assistance whatsoever, though humans may design the process and even initiate it. Interest in the field of molecular and atomic self-assembly is intense because, at present, it appears to be the only path scientists can take to produce the complex nanomachines of the future. Nanoscale self-assembly also offers the promise of someday becoming so inexpensive when carried out in bulk that it could potentially revolutionize the manufacture and design of future technologies, especially in areas such as information science, sensing, biology and medicine.

Nanomanufacturing

The self-assembled nanomachines of tomorrow will be inherently more complex than the machines in use today, and producing them will be equally as involved, requiring new manufacturing paradigms. Broadly speaking, current manufacturing techniques entail placing the component parts of the object to be produced in predetermined positions, according to a blueprint or schematic diagram. Most objects in our everyday lives—our cars, houses, telephones or personal computers—are built this way. Their blueprints not only lay out their structural design, but also take into account the various properties of the materials used and their tolerances, which obviously must form an integral part of the design. For example, a bridge designed with suspension cables stretched to a higher degree of tension than the design and material allow will not be stable. A crane cannot hoist a heavier load than it was designed to hold, given its tolerances. Not only tensile strength or weight must be taken into consideration, but other factors such as center of gravity, hardness, shear strength, air and water permeability, flexibility, or compressibility, must be considered as well in a structural design. At bottom, this form of construction means building from the outside in.

Self-assembly, on the other hand, entails building from the inside out in a process that happens autonomously and is self-driven, so to speak. The process follows naturally from the inherent properties of the atomic or molecular components and the forces that arise between them, such as electrical repulsions or attractions due to the way charge is distributed over their surfaces, producing an intrinsic blueprint. Materials scientists, chemists and others are currently experimenting with various techniques for the design of nanoscale machines and their respective manufacturing processes that are based on the principles of self-assembly. Utilizing this method for fabricating microscopic machines would enable us to construct them in a way that avoids human error and the use of highly elaborate and expensive manufacturing aids like robotics. For some of the simpler, two-dimensional nanostructures, error rates are fairly low, as in the case of self-assembled monolayers (SAMS).

Much of laboratory self-assembly has taken its cues from nature. Indeed, nature can be considered the quintessential architect of self-assembled structures, producing entities as complex as an intricately folded, macro-molecular protein, a snowflake, or a living cell. All these structures are assembled autonomously via interactions between the physical and chemical properties of their component parts. These natural assembly processes, however, are usually so complex that at this point in our understanding of biology, we are still unable to reverse engineer them, in order to deconstruct the various steps in their growth process. Complete control of the intricate interplay of factors at work in atomic self-assembly, like thermodynamics, chemical bonding or electrostatics, still remains outside the grasp of today's researchers. Nevertheless, understanding nature's methods of self-assembly, which are probably vast in number, will be essential to designing tomorrow's nanofabrication processes.

Obervations of existing natural nanomachines in the form of the functional molecular components of living cells—for example, molecules such as mRNA (messenger RNA), the mitochrondrion, or even a biochemical rotory motor that turns the flagella in a bacterium—have much to teach us. The mRNA carries the genetic code, transcribed from DNA, to specialized sites within the cell where the information is translated into proteins. The mitochrondrion generates the power needed by the cell, and the bacterial flagellar motor, a highly organized protein structure within the cell provides the rotary motion for the flagella, small hair-like structures, that propel the bacterium forward in the water. These elegant structures operate so perfectly and so efficiently from the standpoint of thermodynamics, that it's unlikely we will ever be able to replicate them with our laboratory methods; however, this will not preclude us from co-opting some of their properties, in addition to devising de novo methods for nanofabrication.

The study of existing natural self-assembly techniques can also be applied toward the development of an overall strategy for nanomanufacturing, says George Whitesides, Mallinckrodt Professor of Chemistry at Harvard. In his view, self-assembly is the most suitable methodology for carrying out some of the most intricate steps required in nanofabrication, such as those involving atomic-level modification of structure, because it can exploit some of the very sophisticated and well-understood techniques of modern synthetic chemistry. Furthermore, as self-assembly draws heavily from nature, it can even incorporate biological structures directly as components in the final design. In addition, because the final self-assembled structures are those that are thermodynamically the most stable ones, requiring the lowest amount of energy to maintain, they can typically avoid most defects that would be found in manmade objects.

Self-Assembly from the Inside Out

With humans basically considered superfluous in the process, self-assembly takes place according to a unique "blueprint" inherent in the collective properties of the molecular components, which ultimately shape the form and function of the resulting nanostructure. This final product will be the result of interactions such as attraction and repulsion between individual elements, and properties such as shape, surface characteristics, charge, polarization, mass, magnetic dipole, and so on.

Self-assembly requires the component parts to be able to move with respect to each other, so it usually takes place in solution or on the smooth surface of a substrate. The substrate can be modified, for example, by having grooves cut into the surface to guide the assembly process. In order for the final structure to be welldefined and stable, it has to have reached its thermodynamical equilibrium, when all the reactions have reached their completion and the final state is at the lowest, most stable energy state.

The attractive and repulsive forces between the atomic or molecular components in the self-assembly process usually cause the formation of weak bonds between them, such as van der Waals forces, electrostatic or hydrogen bonds. These are called "noncovalent" bonds to differentiate them from the much stronger covalent bonds seen in a typical chemical

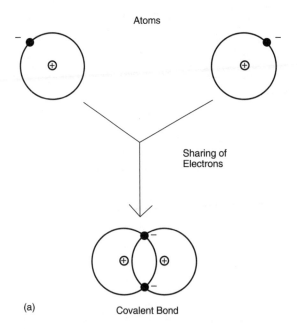

(a)

Figure 6.2
(a) Covalent bonds are formed when atoms share electrons. (b) Noncovalent bonds, such as Van der Waals forces or weak hydrogen bonds, are much weaker than covalent bonds.

compound, as when two hydrogen atoms and an oxygen atom bond together to form a water molecule (figure 6.2). (A covalent bond results when two atoms share the electrons in their outer shells to form a stable molecule. In the water molecule, the two hydrogen atoms each share their outer electron with the six electrons in the outer shell of the oxygen atom, to form the stable compound H_2O. See figure 6.3.)

Noncovalent bonds are only about one-tenth as strong as covalent bonds. For instance, a noncovalent hydrogen bond is an electrostatic bond formed between hydrogen and other atoms that have negatively charged surfaces. Van der Waals forces also cause noncovalent bonding. These are short-range electrostatic forces between uncharged atoms due to the irregular way the electric charge is distributed on their surfaces. All these bonds must also be stable in order for the final nanostructures to remain fixed and permanent.

Biological self-assembly relies upon noncovalent bonding between pre-formed, modular molecular structures, rather than the creation of a single, elaborate structure held together by covalent bonds. The final self-assembled object remains stable, despite its weaker bonds, because it is

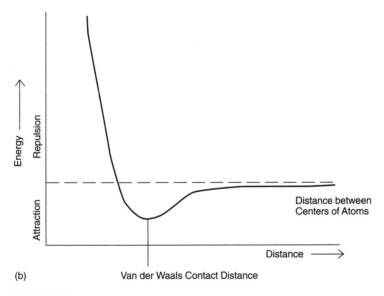

(b)   Van der Waals Contact Distance

Figure 6.2
Continued.

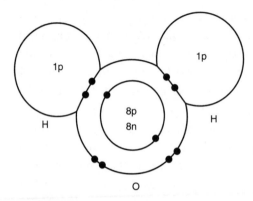

Figure 6.3
The water molecule is composed of two hydrogen atoms, each with one proton forming the nucleus, joined to one oxygen atom, with eight protons and eight neutrons in the nucleus, by covalent bonds.

at a thermodynamic minimum. Faulty substructures are rejected in the process of arriving at the final equilibrium configuration.

Noncovalent bonds can pose problems when used in artificial structures. When in solution, for example, they can sometimes disintegrate because they can't compete with the stronger bonds the individual atoms or molecules in the nanostructure form with the solvent molecules.

Self-Assembly versus Chemical Synthesis

Self-assembly differs somewhat from chemical synthesis, a process that chemists have used for many years in order to fabricate new chemical compounds such as plastic, Lucite, nylon, and pharmaceuticals. On one hand, self-assembly is a completely spontaneous process—even when scientists plan the "spontaneity"—that occurs when atoms under equilibrium conditions form stable, organized structures, primarily with noncovalent bonds, according to their molecular structure and physicochemical properties. Humans do not intervene in the assembly process. On the other hand, chemical synthesis is the process by which chemists derive complex strategies for the step-by-step formation of covalent bonds between molecular species, usually one or several at a time. The process can even generate structures that are far from thermodynamic equilibrium and are unstable. Although it can produce very large molecules, composed of thousands of atoms, chemical synthesis isn't the most practical route to fabricating nanostructures, particularly the larger ones of around 100 nanometers in size. However, it can be used to fabricate the molecular subunits used in modular form to construct larger self-assembled structures.

SAMS and Scaffolds

A simple example of a nano-object formed by self-assembly is a self-assembled monolayer, or SAM. A SAM is a very thin film (1–2 nm) of organic molecules that form a one-unit-thick layer on an adsorbing substrate (*adsorption* is the opposite of *absorption*, where substances are not drawn or sucked into a solid or its surface, but remain attached on the outside). The molecules in a SAM are usually shaped like cylinders. The layer begins to form when atoms at the tip of each cylinder attach themselves to the substrate by bonding with molecules on its surface (figure 6.4). At the cylinder's opposite end there could be any one of a variety of atomic groups, chosen specifically because of their ability to react with another chemical species when positioning a second layer on top of the first. This is accomplished by repeating the process, attaching the tips of the second layer of molecules to the first layer, and so on successively, until a 3D multilayer structure is formed, molecule by molecule. These structures can be imaged by using a scanning tunneling microscope.

One SAM that has been studied extensively is made up of an organic molecule called an alkanethiol—a long hydrocarbon chain with a sulfur atom on one end—on top of a metallic gold substrate. This particular SAM design has received a lot of technical interest because it is fairly easy and inexpensive to make, and could potentially be used in many different technological applications. For instance, in the process of chemically etching a pattern on a metallic surface, the SAM could serve as a sort of "mask" or resist, which would protect sections of the underlying surface from being touched by the chemicals doing the etching. This could be used to make nano-sized features for wiring microscopic integrated circuits used in future information technology.

Chemists Christine Keating and Tom Mallouk at Pennslyvania State University have designed a similar self-assembled object. With the objective of creating "self-wiring" electrical nanocircuits, they have used DNA

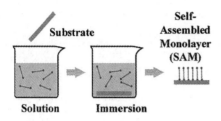

Figure 6.4
The formation of a self-assembled monolayer (SAM).

tags to position tiny gold wires, 200 nm wide and 6,000 nm long, onto a gold surface. The DNA tag on the wire binds to a complementary piece of DNA attached to the surface, like a hook and an eye, thus guiding the wire into place. Keating has used this same technique to create more complex patterns linking the wires together, and hopes ultimately create a nano-sized integrated electrical circuit (figure 6.5).

Another example of a promising self-assembled object is the microscopic liposome, a biomimetic spherical sac enclosed in an artificial membrane that can be used to transport relatively toxic drugs into diseased cells inside the body without affecting other parts of the body. For example, in treating cancerous organs, a liposome containing a drug called methotrexate is injected into the body. The diseased organ is heated above body temperature, so that when the liposome passes through it, the membrane melts and the drug is released. To further improve the specific binding properties of a drug-carrying liposome to a target cell (such as a tumor cell), specific molecules (antibodies, proteins, peptides, etc.) are attached on their surface.

An MIT bioengineering professor, Shuguang Zhang, has produced some self-assembled biostructures that show great promise for therapeutic purposes. Zhang showed that certain protein fragments, called peptides, can be manipulated into self-assembling into forms like fibers, sheets, tubules, and thin layers, thus creating a new class of natural structures that potentially have many useful applications. For example, Zhang and his group have experimented using them as scaffolds for tissue engineering to repair damaged cartilage in the joints.

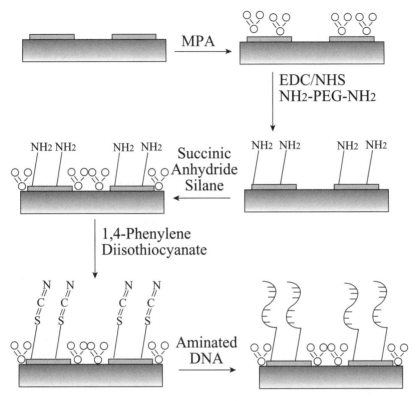

Figure 6.5
Linking gold nanowires to surfaces with DNA. (Diagram courtesy of Christine Keating, Pennsylvania State University's Department of Chemistry)

Self-Assembly and Biocomputation

Developing proficiency with molecular self-assembly represents a critical step in the advancement of the field of nanotechnology, bringing many of tomorrow's nanomachines and nano-objects closer to our grasp. The rationale for including it in a book on biocomputation, however, may not be obvious and require further explanation. On the level of basic physics and chemistry, self-assembly is a process that involves interactions between groups of atoms or molecules based on their physical or chemical properties, and other variables such as temperature or pressure. The chemical properties of a certain solvent, for example, at a certain temperature, will cause a dissolved solid to crystallize out from the liquid. The electrical charge distributed over the surface of two atoms will either cause them to form one type of bond or another, or else repel each other. These interactions are common processes in molecular self-assembly, and form part of the wide repertory of techniques that scientists would like to incorporate in their toolbox for the design and manufacture of organized nanostructures.

In nanofabrication, a scientist must be explicitly aware of the individual steps involved in the sequence of physical or chemical reactions that constitute the self-assembly process, so he or she can carefully design and then implement them experimentally in the lab. In fact, we could compare the instruction set laid out in the nanostructures blueprint for assembly to a computer algorithm, and similarly, on another level, the self-assembly process itself to a computation, as the latter is simply a series of explicit instructions, written in code, that tells a computer specifically how to solve a problem.

Some scientists in this highly multidisciplinary field believe the use of concepts and language borrowed from computer science—more than a just an interesting metaphor for the self-assembly process, can open up a new and different perspective on these phenomena. They see the self-assembly process as actually constituting a computation, as it were, when viewed from a higher, or different level of abstraction. (The "hierarchy of differing levels of abstraction" argument is not uncommon in multidisciplinary fields of science. For example, when referring to brain activity, a biophysicist may talk about ions crossing cell membranes in a neuron, thus creating an electrical potential difference across the cell; a neurobiologist would talk about neural activity in a brain area like the amygdala, whereas a psychologist might refer to a person experiencing the emotional feelings of love, but they would all be describing the same phenomena from their own point of view. The hierarchy of differing levels of abstraction for this phenomenon-atom, molecule, ion, cell, neuron, neural network, amygdala, brain activity, emotion, love—each with its corresponding specialized body of learning and language, enables one to simply choose the level that works best for their purposes.) Seen in this frame of reference, viewing self-assembly as an algorithmic or computational process means reverting to another level in the hierarchy of abstractions as one scales up from physics and chemistry to the field of information processing.

Moreover, it's possible that this approach may even bring us a step closer to realizing tomorrow's nanofabrication processes. Advocates claim that likening the steps in the self-assembly process to algorithms could enable a greater degree of specificity in the instructional blueprint, particularly if the structure is highly complex. In addition, it could reveal hidden connections with the world of information processing and mathematics, which could also prove useful in creating regular patterns in nanostructural design.

Skeptics, however, claim that using computational concepts and terminology to characterize self-assembly is really just a question of semantics, and that while the idea of computation as metaphor might have its appeal, it's not terribly useful as a description.

Self-Assembly with DNA

One subset of the field of self-assembly may lend itself more naturally to a description in the language of computation: DNA self-assembly. Leonard Adleman's original work at the University of Southern California using DNA strands to compute the solution to the Hamiltonian Path Problem was for all intents and purposes a *computation* in that Adleman had worked out the precise sequence of base pairs needed to represent each city and each flight between them. The base pairs constituted a code, and manipulating these codes embedded in the DNA strands was in effect the implementation of an algorithm. This may be one reason why many computer scientists have been attracted to the multidisciplinary field of DNA computation. Other factors, such as the abundance and availability of well-known recombinant DNA lab techniques for manipulating DNA, have also helped the field to grow.

New York University chemist Ned Seeman has labored for over two decades developing extremely sophisticated lab techniques for implementing nanotechnology with strands of DNA, and has shown that highly elaborate DNA structures, such as branched junctions, knots, and complex geometrical shapes and patterns can routinely be designed in vitro (figure 6.6). These elaborate manipulations require such a high degree of skill and dexterity, and often can be so difficult to produce that the yields are typically very small, claims Seeman. However, he believes the technology will get easier as more people begin working in the field,

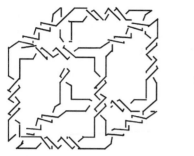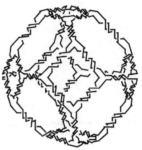

Figure 6.6
Stick figures constructed from DNA branched junctions, by Ned Seeman's lab at New York University: (left) a stick cube and (right) a stick truncated octahedron. The drawings show that each edge of the two figures contains two turns of double helical DNA. The twisting is confined to the central portion of each edge for clarity, but it actually extends from vertex to vertex. (Courtesy of Ned Seeman)

and knowledge about how to avoid fabrication errors becomes more widely shared.

Caltech scientist Erik Winfree has combined Seeman's work with knowledge of Adleman's programming techniques for DNA computation to design and study organized geometrical arrays of DNA known as tilings (the patterns that emerge can often resemble those in bathroom tiles, hence the term "tilings"). The construction, via self-assembly, of these ordered patterns of DNA structures has shown considerable promise not only as a means for nanofabrication—serving as templates or scaffolds for more elaborate structures—but also because they can be seen to implement a form of DNA computation. Winfree has been able to map the mathematics of "Wang tiles"—a mathematical problem posed in the 1960s by Chinese mathematician Hao Wang about the logic underlying the placement of geometric tiles in a pattern—onto Seeman's knotted DNA structures. In the process, he has also shown that this type of assembly emulates the operation of a universal Turing machine. Winfree believes his work will help scientists better understand the algorithmic design, stochastic behavior and error control involved in future self-assembled molecular systems.

Smaller and Smaller

Mastering the techniques of molecular self-assembly and devising new ones will be critical to future advances in nanotechnology, as it is one of the principal methods for fabricating objects on this scale. Future nanostructures won't be built with tiny robotic pincers in a miniaturized mechanical assembly line that will grasp the molecules and hoist them into place, as some may believe. These atomic and molecular LEGOS will have to self-assemble into organized, predesigned structures and that involves manipulation, control and production at molecular dimensions, without any mediation by humans. The corresponding nanoscale design, verification and testing procedures will also be based on this methodology.

Over the next few decades, the current thrust for smaller and smaller devices, culminating in a new manufacturing paradigm of nanotechnology, will have a profound impact on the world. Learning how to manipulate and control matter at the atomic scale will have far-reaching implications for healthcare, the environment, sustainable development, industry, manufacturing, defense, business, trade, entertainment and other fields of activity affecting the fabric of our daily lives. Developing a greater understanding and command of the principles and techniques of self-assembly, and learning how to implement them *en masse* at very low cost, will almost certainly usher in widespread changes to our society, possibly even as sweeping as those brought by the industrial revolution or today's information age.

7

Amorphous Computing

Listen to a man of experience: thou will learn more from Nature than in books.
—St. Bernard, *Epistle 106*, twelfth century

Even as the first digital computers were being designed in the 1940s, the founders of computer science were concerned about the inherent limitations imposed by the need for extreme precision and reliability in machines that processed information. In his work on automata, John von Neumann had observed that in a computer, even one small error at any stage of a billion-step process could completely invalidate the end result, or even disable the machine. In contrast, he noted, natural organisms had a remarkable ability to handle errors and malfunctions. Even when damaged, most living things can usually continue to function because of their built-in self-healing capacity and redundancy (for example, if you break an arm or contract a non-lethal illness, in most cases, your entire system doesn't shut down). In von Neumann's words,

With our artificial automata we are moving much more in the dark than nature appears to be with its organisms. We are, and apparently, at least at present, have to be much more "scared" by the occurrence of an isolated error and by the malfunction which must be behind it. Our behavior is clearly that of over-caution, generated by ignorance.[1]

Tackling Fault Tolerance

Fifty years of experience with computers have shown that von Neumann may have been too pessimistic. He believed that expanding from computers with a few million logical elements to those with several billion would require major advances in computation theory. However, we

continue to build our computing machines today on the same foundations von Neumann originally laid down and only now are learning to
look at the issue of errors and reliability in a different way, constructing
computers that are more tolerant to faults than before. Perhaps more
important, we are finally beginning to heed von Neumann's admonition
to look to natural organisms for clues.

At MIT's Artificial Intelligence Lab, senior scientists Hal Abelson, Tom
Knight, and Gerry Sussman, together with their graduate students, have
taken up von Neumann's challenge in earnest, turning to biology for
inspiration to address the problem of errors and unreliability in
computing. They have dubbed their efforts in this area "amorphous
computing."

A preoccupation with rigor and precision has been a common thread
throughout the work of Gerry Sussman. He firmly believes that the time
has come for computers to show greater flexibility and more tolerance
to defects. "Computing is in real trouble," he claims. "Systems, as
currently engineered, are so brittle and fragile, that minor changes in
requirements entail large changes in structure and configuration. In addition, our current computational systems are unreasonably reliant on the
implementation being correct. Just imagine what cutting a random wire
in your computer would do, for example." Sussman calls for a new set
of engineering principles for building flexible, redundant, and efficient
computing systems that are "tolerant to bugs, and adaptable to changing conditions."

One way to tackle fault tolerance, Sussman suggests, is to design
redundancy into the system, both in terms of hardware and software,
analogous to the principles found in biolological organisms. "Just as a
colony of cells cooperate to form a multicellular organism under the
direction of a shared genetic program, or a swarm of bees work together
to build a hive, or humans group together to build cities and towns,
without any one element being indispensible," a program should be able
to "prescribe the system as a cooperating combination of redundant
processes," argues Sussman. He underlines his point by citing an example
in biology: "The cell has two ways or two separate pathways, for extracting energy from sugar—one involving oxygen and the other without it.
Depending on how available oxygen is, it may use either of the two path-

ways. The system adapts and makes the optimal choice, depending on the availability of resources and the environment."[2]

Amorphous Computing

The goal of the MIT group's amorphous computing project is to design a system and accompanying algorithms that embody these principles, while also efficiently processing information. The team not only believes this goal is possible, but that the time is ripe for it. This is due, they assert, to the tremendous advances in the last few decades in our understanding of biology, together with the current availability of cheap microprocessors. These technologies will make it possible to assemble "amorphous" systems composed of thousands of information processing units at almost no cost, assuming that all the units need not work correctly and there is no need to manufacture precise geometrical arrangements of the units or precise electrical interconnections among them.

As electronic devices and machines become smaller and smaller, reaching sizes down to thousandths of an inch (these devices are called microelectromechanical systems, or MEMS), they will cost increasingly less to manufacture and become more ubiquitous in our everyday electronic devices, including computers.

Microelectromechanical systems are considered the next step in the silicon revolution. The number of transistors incorporated onto a piece of silicon has grown exponentially over the years, leading to small, high performance, inexpensive integrated circuits (ICs). Many observers feel that the next phase will be for ICs to incorporate new types of functionality—structures that will enable it to think, sense, and communicate. These microelectromechanical systems represent a major technological stride forward.

One mode of carrying out the amorphous computing paradigm, according to the MIT group, is via "computational particles" composed of integrated microsensors, actuators, or advanced communications devices on the same chip. These particles could be mixed with bulk materials, such as paints, gels, or concrete, at a low cost to produce a distributed sensor network inside the medium itself. One could coat bridges or buildings with "smart paint" that could sense data, act on it, and communicate its actions to the outside world. For example, the paint could report on traffic and wind loads, or even heal small cracks by shifting the material around. A smart paint coating on a wall could sense vibrations, monitor the premises for intruders, or cancel out surrounding noise (figure 7.1).

Moreover, advances in scientific instrumentation in the twentieth century, such as x-ray diffraction and electron microscopy, have enabled us to probe much more deeply into the fundamental biochemical mechanisms of the cell at the molecular level than we could have before with a simple light microscope. This has provided new concepts for cellular organization, structure, and function. The study of the molecular organization of the cell has probably had the greatest impact on biology in the twentieth century, and also caused the convergence of many scientific disciplines—physics, chemistry, and statistics, for example—in order to gain a better understanding of life processes.

As a principle theorist behind amorphous computing, Hal Abelson has defined some of the basic rules required for information processing with computational elements such as these, in the kind of adaptable, biolog-

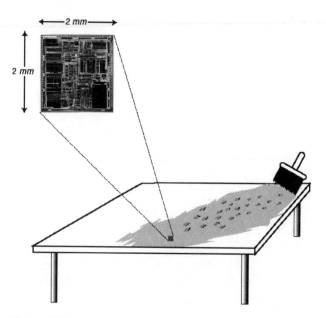

Figure 7.1
"Smart" paint. (Image courtesy of William Butera, MIT Media Lab)

ically inspired model the group envisions. "First, errors and faulty components must be allowed and won't disrupt the information processing or affect the end result; and second, there will be no need for fixed, geometrical arrangements of the units or precise interconnections among them."

He continues:

We've seen some of these properties before, for example, in studies of self-organizing systems. But the phenomena there are largely resultant rather than purposive or by design. One area we look to for suggestions is developmental biology, where the processes of cell differentiation and morphogenesis appear to perform a kind of information processing in order to reach the desired end state. Many details of these low-level cellular mechanisms are already known—such as diffusion, gradient following, growing points, regularization, dominance, activation and inhibition—although the overall organizing principles in developmental biology are still elusive.[3]

The group has experimented with several different kinds of computing hardware for implementing amorphous computing. In the standard logic circuit, the information processing depends on turning on or off an electrical current, which corresponds to digital bits of zero's and one's. The electrical wires that are laid down on a substrate are the physical means that connect the information processing elements. The algorithm used fixes the configuration they form, which doesn't change in the course of the computation; there is no flexibility, adaptation, or redundancy built into the physical system.

In the case of amorphous computing, the substrate needed as a computing platform ultimately must enable a system of "cooperating combinations of redundant processes," as Sussman describes it.[4] One method for implementing an amorphous computer relies on a system of irregularly placed computing elements, whose individual function is allowed to change over time (i.e., they are not synchronized). These units are programmed identically, and incorporate hardware for booting up, for networking with neighbors, and generating random numbers.

The group's experimental set-up for amorphous computing is composed of microchips that incorporate a microprocessor and a short-range spread-spectrum radio transceiver that emits and receives radio waves. These waves connect the chips, taking the place of electrical circuits in regular microchips with wire interconnects (see figure 7.2). If one of the computational elements breaks down and stops working, the previous

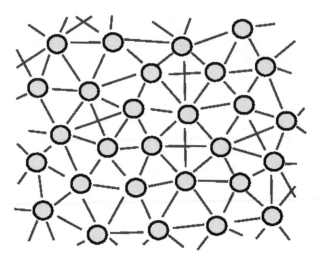

Figure.7.2
Idealized diagram of amorphous computing setup, where circles represent individual computional "elements," connected by radio waves instead of wires.

element will simply ignore it and continue to transmit the signal to its next nearest neighbor, thus fulfilling the requirement that errors shouldn't matter.

If a large number of these particles (up to a thousand) is "sprinkled" at random on a two-dimensional substrate, called a "computational lawn," that provides the power, each particle can communicate with its neighbors via the radio waves. In this scheme, power and memory required would be modest, and each computational particle would be so inexpensive that a dollar could buy millions (see figure 7.3).

Teramac

Although it seems plausible that such untraditional hardware could work, the issue of how to program such a system still remains. The challenge posed here is to obtain organized, prespecified behavior from the cooperation of vast numbers of unreliable elements that are interconnected in a priori unknown, irregular, and constantly changing ways.

One approach to programming such systems is to have it begin with a self-discovery and diagnosis phase, where the elements test each other to discover which ones are operational or not, and how they are inter-

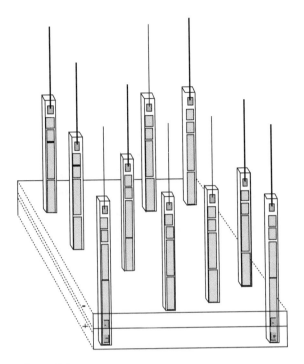

Figure 7.3
Sketch of computational elements embedded in a "computational lawn" that provides the power to connect the elements via radio waves. (Image courtesy of Tom Knight, MIT AI Lab)

connected. The system can then reconfigure itself and its communication paths to avoid the broken parts and compensate for irregular interconnections. The Teramac computer, a massively parallel computer constructed from defective chips by computer scientist Phil Kuekes and others at Hewlett-Packard Laboratories (HP) in Palo Alto, California, uses this method. Reporting in *Science*, Kuekes, together with colleagues from HP Labs and UCLA chemist James Heath, explains how Teramac, the largest defect-tolerant computer built so far, was able to perform certain tasks faster than HP's top of the line work stations, even though the hardware had over two hundred thousand defects.[5] Teramac, with 10^6 gates operating at 1 MHz, was built from conventional chips, called Field programmable gate arrays or FGPAs. These are look-up tables connected by a large number of wires and switches organized to form crossbars, allowing one to connect any input with any output. (Field programmable gate arrays substitute memory for logic whenever they can. With such a large computer, it made sense to utilize this method and store as many intermediate results as possible, looking them up when needed.) The high degree of Teramac's connectivity makes it possible to access most of the good components and avoid the defective ones. (According to Abelson, the hardest thing about Teramac was getting HP's meticulous Japanese suppliers to send faulty chips, which they were loath to do.)

Teramac's fault-tolerant architecture was planned as a prototype framework for designing nanoscale computers whose logic elements would be assembled by chemical processes. The underlying assumption is that these chemically synthesized electronic components would not be produced with a 100 percent yield, and so the Teramac architecture could continue to function perfectly, working its way around the faulty elements, even while locating and tagging the defective parts. In addition, the Teramac structure would enable a chemically assembled computer to reproduce any arbitrarily complex program. The HP researchers found, to their surprise, that the compiling time for new logic configurations was directly proportional to the number of resources used, and that the execution time was remarkably fast (about a hundred times faster than a top-of-the-line work station—Teramac dates from 1998).

In Teramac, the software compensated for the defects in the hardware—it took about a week to be able to locate, map, and catalog the faulty chips (it turns out that of the more than 7 million total chips in Teramac, 3 percent were defective). The software then commenced to rewire the computer to avoid the problem areas. Because of this capability, Teramac could optimize its hardware, first for one task and then for another.

Although both Teramac and amorphous computing models are purposely designed to deal with parts failure, they each approach the matter differently. Explains Abelson,

We and HP are on the opposite side of the spectrum when it comes to fault tolerance. HP would say that their way with Teramac is to learn to build machines that really do conventional, reliable computation, and then, before any information processing begins, if there are faults, the computer can them diagnose and fix themselves or else route the processing so that it avoids any faulty chips. We, on the other hand, would say that biological systems don't work like that. In this case, and with amorphous computing, you have fundamental algorithms that are independent of whether or not the computational units are completely functional and reliable. Ultimately, I think that both kinds of systems will end up being very useful.[6]

According to MIT's Tom Knight,

While Teramac represents a static failure model, amorphous computing actually rethinks the way we compute things. Teramac starts out with an imperfect substrate, finds out which are the defective components, fixes or else avoids them in processing, and then superimposes that model on top of a perfect computer, which they'll use with very conventional sorts of algorithms. The amorphous computing approach deals not only with static, but with dynamic failures as well—its fault tolerance comes from the choice of algorithms. The idea is to choose algorithms that are inherently robust to point failures, so if one element doesn't work, the algorithm readjusts itself and continues as if nothing happened. In amorphous computing, you embrace the imperfections, and deal with them not only at the hardware level, but also at the algorithm design and implementation levels.[7]

This means that programming would require new organizing principles and techniques that do not require precise control over the arrangement of the individual computing elements. Here, as von Neumann hinted, biology served as a rich source of metaphor for new amorphous software methodologies.

Amorphous Software

The phenomenon of morphogenesis—the growth of specialized forms in cellular organisms such as the way a hand or ear develops in an early embryo from an initial fertilized egg—demonstrates that well-defined shapes and forms can evolve out of the interaction of cells under the control of a genetic program, even though the precise arrangements in the form and numbers of the individual cells are highly variable. Inspired by these process, the MIT investigators are developing algorithms by which an amorphous system can self-organize and differentiate itself into the equivalent of "discrete tissues, organs, and systems of organs that perform specific functions."

The group is experimenting with such ideas, simulating a group of cells by using a large number of their computational particles distributed in a random but homogeneous manner on a surface. The "software" they use is called Growing Point Language (GPL), developed by a former student of Abelsen, Daniel Coore. Growing Point Language is modeled after the biological phenomena of growth, inhibition, and tropism (e.g., "phototropism" is a phenomenon that causes plants to grow in the direction of a light source).

In GPL, the computational elements are placed at random in a computational lawn. The units' internal clocks are not in phase; some units may be faulty, some may be affected by the environment, and all are limited only to interactions with their nearest neighbors. Each unit has some memory, is programmable, can store local state information, and can also generate numbers at random. The units communicate via short-distance radio waves over a distance R, which is small compared with the dimensions of the entire lawn, but large compared with the size of the particles. The way they communicate is similar to the way broadcast networks with overlapping regions communicate.

To get a sense of a simple GPL program, one could imagine the way a radio wave would diffuse throughout the units. An initial particle, on some outside cue, broadcasts a radio wave message to each of its neighbors. The message contains a "hop count," which estimates the number of hops it has to make in order to reach a given unit. On getting there, the cell stores and rebroadcasts the hop count number, augmented by one. In this way, the hop count gives a rough estimate of the distance of each cell from the origin. A hop count of seven will mean the message has jumped seven units from the origin, or that the final unit is 7R away from the origin.

One particular GPL algorithm mirrors the biological phenomena of growth and inhibition. Here, two particles, X and Y, each produce a diffusion wave, but the wave from Y is relayed only by particles that have not seen the wave from X. In other words, X generates a wave that inhibits the growth of Y. A similar program might make the wave from Y propagate only via those cells that are in the neighborhood closest to X, as measured by the X-wave. Continuing the biological metaphor, one could say that the region growing from Y exhibits a "tropism" attracting it toward X. Both these GPL diffusion wave programs work well with amorphous computing because they don't depend on where the units are placed, or whether any one of them is malfunctioning, as long as there is large enough density of units on the substrate.

The main concepts of Coore's GPL is the *growing point*, which represents a locus of activity that describes a path through the elements on the lawn. The growing point propagates throughout the system by transferring its activity from neighbor to neighbor, causing each element that it reaches to change its behavior, in morphogenesis-like fashion. The trajectory of the growing point is controlled by signals carried throughout the system from other differentiated cells, much like the *tropism* above.

Growing Point Language aims to emulate the way the cells of an embryo specialize and develop under the control of a common genetic program housed in each cell. From embryo to developing fetus, the cells eventually grow and differentiate, ultimately to form a hand, a foot, or a finger. Growing Point Language's developer, Dan Coore, was interested in exploring this mechanism and seeing how cells differentiate, despite carrying the same genetic code; he wanted his amorphous program to mimic this. Coore was able to prove that with GPL, one could specify a priori and generate any kind of 2D pattern, no matter how complex. As an example of a complex pattern, he used the interconnection pattern of an arbitrary electrical circuit. However, according to Abelson and Sussman, their "amorphous" vision goes far beyond that to encompass a future technology where—if amorphous algorithms can be programmed to construct detailed patterns, and if, one day, living cells can be programmed to carry out specific cellular functions—then a system of cells could be collectively programmed to construct electronic devices, for example, or carry out a biomedical function.

Figure 7.4 shows an example of the pattern generated by Coore, a chain of CMOS (complementary metal oxide semiconducter, a type of logic circuit) inverters, where the different shaded regions represent structures in the different layers of standard CMOS technology: metal, polysilicon, and diffusion. In principle one could reduce any prespecified electrical circuit pattern to an amorphous system, but whether this would be a sensible way to use the power of amorphous computing remains to be seen. MIT postdoc Radhika Nagpal has developed a method of programing amorphous "sheets" to fold themselves into pre-determined shapes, as a precursor to an "intelligent" material, and has also developed amorphous algorithms that attempt to self-heal.

As for future directions, Tom Knight is optimistic. "With amorphous computing, we'll learn a lot more about how to program computers than we do now. We'll learn how to build better algorithms and organizing principles that can cope with faults differently from the standard von

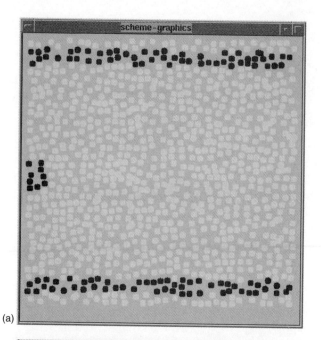

(a)

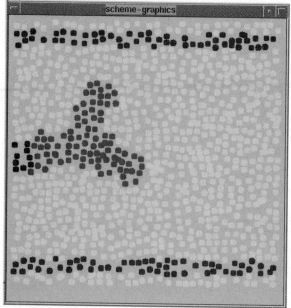

(b)

Figure 7.4
Growing Point Language (GPL) used to "program" an amorphous computing setup to generate the interconnection pattern of an arbitrary electrical circuit. The dark points starting in the middle of the left side of the image in (a) represent the evolution of the pattern in GPL to (b) and (c). (Image courtesy of Tom Knight, MIT AI Lab)

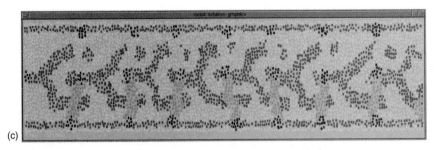

Figure 7.4
Continued.

Neumann way of duplicating and triplicating things to achieve redundancy."[8]

The overarching goal of amorphous computing is to draw from biology to help create a new branch of computer science and engineering concerned with orchestrating the use of masses of bulk computational elements to solve problems. In addition, this could provide the enabling technology for bulk "intelligent" engineering materials that incorporate computational elements, sensors, and effectors. Additional exploration of amorphous computing may uncover principles that will not only further the exploitation of biological metaphors for information processing, but could well become a dominant factor in the next generation of computing and advanced engineering systems.

8

Computer Immune Systems

Our remedies oft in ourselves do lie.
—Shakespeare, *All's Well That Ends Well*, I, i

Information technology has become indispensable to our lives and vital to the functioning of our societies, so much so that its vulnerability to failure or attack has become a subject of increasing concern. Computer systems form an integral part of our critical infrastructure, such as electric power grids, nuclear power plants, dams, air traffic control systems, and large financial institutions. Parts of our food and energy distribution systems are now computerized, and computers are now embedded in most of our devices and networked into larger systems. This means that any large-scale computer system failure, virus or cyber attack could have enormous, perhaps devastating consequences, costing billions of dollars, and possibly even lives.

New computer viruses are being written every day, often overwhelming the experts who try to find cures. What's more, the growth of networked computer systems and interoperability cause viruses to spread much more easily.

This high degree of vulnerability of our computing systems has caused significant growth of the discipline of computer and network security, where students, IT professionals, and researchers are seeking more effective methods for monitoring and protecting these systems, as well as defending and restoring them if attacked. Although new procedures have been developed that can be applied to computer systems and networks to help ensure the availability, integrity, and confidentiality of the information they manage, the problem of how to secure computer systems against unwanted intruders still remains a complex and highly challenging one. No one existing method is foolproof, nor offers complete, permanent protection.

Can Biology Suggest a Better Way?

One approach to protecting computer systems finds its inspiration in the human immune system. There is probably no system, living or manmade, that can protect itself as effectively, on so many levels, and from so many different diseases and infections as the human immune system. This remarkable complex is composed of a network of organs—highly specialized cells and a circulatory system separate from blood vessels—which all work in tandem to maintain the body's health. This system has evolved over hundreds of millions of years in the course of responding to invasions by pathogens (any disease-causing microorganism) that tried to infect our genetic ancestors. In fact, there are distinct similarities between the human immune system and that of the most primitive mammals, going back five hundred million years on the evolutionary ladder.

The targets of the immune system are infectious organisms—bacteria, parasites, viruses, and so on—that can inflict serious, even lethal harm once inside the body. Its ability to defend the body against disease is based on its capacity to distinguish between *self* and *nonself,* the former being those cells that are recognizably integral parts of the body and essential to its life and health, whereas the latter are outside or foreign agents that could cause it harm.

Inside the Immune System

The notion of *self*, in the context of a living immune system, comes from the fact that all the body's cells carry distinctive molecules that distinguish them as *self* (figure 8.1a). Normally the body's defenses recognize this and do not attack these cells; in a healthy system, immune cells coexist peaceably with other cells in the body in a state known as *self-tolerance*. In fact, when this balance goes awry, the immune system begins to attack the body's own cells, resulting in what's called an auto-immune disease, such as lupus erythematosus or rheumatoid arthritis.

Foreign bodies also carry distinctive markers enabling the body to recognize them as *nonself*. These characteristic shapes are called *epitopes*, which protrude from the cell's surfaces (figure 8.1b). One of the remarkable aspects of the immune system is its ability to recognize billions of different kinds of nonself molecules—possibly even up to 10^{16}. To do this, they must differentiate them from the body's own cells, of which there are about two hundred different types. This makes the cell recognition task highly targeted and specific.

The biological immune system responds to an attack by producing antibodies that fight the invader in question by identifying and counteracting it. Any substance capable of triggering an immune response is known as an *antigen*. An antigen can be a bacterium or a virus, or even a portion or product of one of these organisms. Tissues or cells from another individual can also be antigens, explaining why the body often rejects transplanted organs as foreign.

In addition to its specificity and extensive range of responses to invading pathogens, another remarkable property of the human immune system is that it functions on many levels, each one reinforcing the other (figure 8.2). The outermost barrier against disease and germs consists of the skin and mucous membranes, such as in the nose. If invaders do manage to penetrate this barrier, a second layer of protection lies in bodily properties such as its normal operating temperature (not too much below or above 98.6°F) or degree of acidity (pH level) that provide environments where many microbes can't survive.

Any pathogen that does manage to penetrate the outer two levels of protection is dealt with on a cellular level by the body's inner immune system, which has an innate and adaptive part. The innate immune system is composed of nonspecific elements that can distinguish a general class of foreign agents or organisms, but are unable to recognize a specific individual pathogen. They always respond to pathogens in the same way, even after repeated exposures to it, rather than adapting and improving their effectiveness against previously encountered microbes.

The best known element in the innate immune system is the *macrophage*, known mostly by its moniker, "white cell." These cells circulate throughout the body and are the major killers of germs and microbes. In addition to being critical in the destruction of

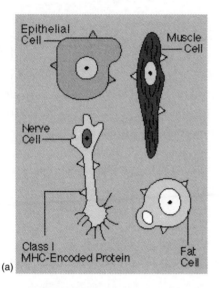

(a)

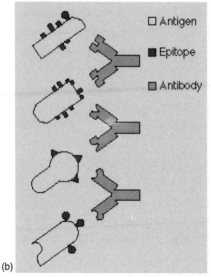

(b)

Figure 8.1
(a) Every cell in the body, like those shown here, has a marker that distinguishes it as "self," so it is not attacked by the body's defenses. (b) Foreign pathogens or antigens carry markers, called epitopes, that distinguish them as "nonself." (Source: National Cancer Institute of the National Institutes of Health)

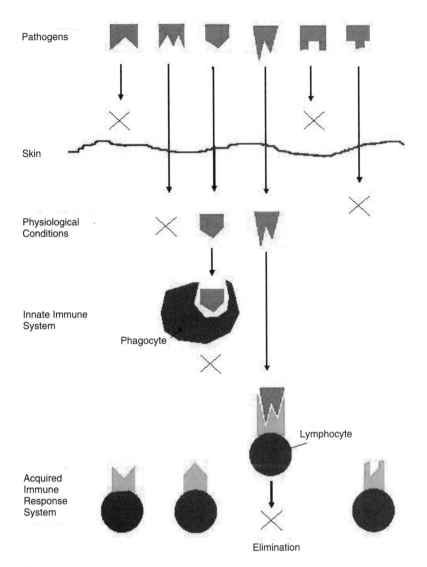

Pathogens

Skin

Physiological
Conditions

Innate Immune
System

Phagocyte

Lymphocyte

Acquired
Immune
Response
System

Elimination

Figure 8.2
The body's immune system has a multilayered defense. (Image courtesy of
Stephanie Forrest)

dangerous organisms and cancerous tissues, they are also key facilitators of communication, via chemical means, between immune system elements.

The adaptive or acquired immune system is the more complex of the two systems, and the least understood by scientists. Its ability to create immunity to disease is built up throughout the lifetime of the organism. Two central features of the adaptive immune system are its specificity—in that it can distinguish self cells from nonself cells and one nonself cell from another—and its memory, a property that enables the body to become immunized. Once an organism has been exposed to a given antigen, such as the chicken pox virus, it will remember this enemy for an extended period of time, continuing to protect the body against it.

For example, vaccines usually consist of a dead or much weakened form of the antigen it intends to protect the body against, such as a flu shot or smallpox vaccine. The weak antigen will call forth a specific immune response to its type, thus creating a defense against the stronger antigen for a given length of time.

The adaptive immune system is made up mostly of white blood cells called *lymphocytes*—either T cells or B cells—that circulate through the body looking to find and eliminate invaders (nonself cells). A lymphocyte recognizes an antigen when certain molecular bonds are formed between receptors on the surface of the lymphocyte and the antigen. However, the detection mechanism is not perfect because a lymphocyte can bind with several different kinds of pathogens.

A wide variety of lymphocyte receptors are required so that they can bind to the billions of different kinds of foreign agents. This diversity stems in part from the body's ability to generate many random types of lymphocyte receptors. Among these, the reason there are not more generated that are capable of destroying self cells—and thus causing an autoimmune disease such as AIDS—is because when lymphocytes mature in the body's thymus gland, almost all of the self-recognizing kinds are destroyed before becoming part of the immune system.

Nonetheless, there is still not enough diversity within the range of random lymphocytes generated to cover the 10^{16} different types of pathogens. The body handles this in several ways. One way is to have a continual turnover in lymphocytes, as they typically live only a few days, and are constantly being replenished with new ones. This dynamic form of protection means that the longer a pathogen is in the body, the more likely it will be recognized by one of the broad range of existing lymphocytes.

A second form of protection results from the immune system's ability to learn and remember a given pathogen-specific response. If the body encounters a pathogen for the first time, it will "learn" the structure of this pathogen and "remember" it by developing a set of lymphocytes geared specifically to recognize it, as part of the Darwinian process of natural selection. These lymphocytes will then reproduce in great number, which accelerates the detection and elimination process. This is why when a child develops measles, its immunity to the disease lasts for a prolonged period of time.

Despite the exquisite sensitivity and versatility that evolution has bestowed upon the vertebrate immune system, its coverage is still not perfect. There will be pathogens, such as the Ebola virus, that it can succumb to. However, not everyone's immune system is the same, and this great diversity across the species helps to ensure its survival.

Digital Immunology

Because our immune system has evolved and adapted over the centuries, improving and perfecting its ability to defend the organism against disease with each new generation, it provides a remarkably good model for creating a computer immune system. As IT researchers have discovered, the challenge has been in finding appropriate analogues between protein-based biological structures and digital ones. For either standalone computers or several computers connected by a network, there are several properties of the biological immune system that seem especially suited to a digital immune system. These are:

Distributability. The biological immune system is a completely distributed system that functions throughout the entire body and is not subject to a central command site. Analogously, an ideal digital immune system would be uniformly distributed throughout a single computer system or a series of networked computers.

Fault tolerance. The living immune system is highly fault tolerant or redundant, in that individual elements can fail without affecting the system's functioning as a whole.

Memory.   The biological system can be trained to retain the memory of its response to a particular pathogen, extending an immunity over time.

Autonomy.   The human immune system is autonomous, in that it functions without our being aware of it or consciously willing it to do so. Likewise, a digital version would not have to be controlled by the user.

Diversity.   Just as there is genetic diversity among individuals in a given species, and a diverse set of mechanisms for recognizing and destroying antigens within the mammalian immune systems, there is also a need for diversity given the variety of different computer viruses a computer immune system would have to respond to.

Multi-layered.   The living immune systems has multiple layers that are mutually reinforcing; a similar capability could be built into a computer protection system.

Dynamic coverage.   Our body is continually replenishing its store of lymphocytes and randomly generating new ones to increase its breadth of coverage. In computers, this would mean that new methods would have to be devised for dealing with new security problems as they emerged.

Novelty detection.   Our bodies can detect pathogens it has never previously encountered before; the same should exist digitally.

More than any of these properties, though, it is critical that a digital immune system have the ability to distinguish between self and nonself.

Variations on a Theme

The idea of approaching the problem of computer security using an immune system model was developed independently by computer scientists Stephanie Forrest at the University of New Mexico, in Albuquerque, and Jeffrey Kephart of IBM's Thomas J. Watson Research Center in Yorktown Heights, New York, in the early 1990s. Though both scientists sought to incorporate the same basic principles into their systems, their versions differ, with Kephart's being geared more toward virus protection in networked computers rather than stand-alone systems.

Forrest believes a biological approach to security is more in sync with the uncontrolled, dynamic, and open environments in which computers currently operate, as users are constantly being added to or removed from a networked system, and programs and system configurations are constantly being changed. Standard security measures such as encryption or user access control, says Forrest, are important and can complement the immune model metaphor; however, they are not sufficient by themselves. For example, in the case of computer viruses, new species are being generated every day, making commercial antivirus software quickly outdated, and often leaving the experts responsible for analyzing and finding cures for them unable to keep up with the changes.

Forrest found the immune system model well suited to dealing with the range of computer security problems encountered today. With her research group at the University of New Mexico, she has applied immune principles to three specific areas of computer security: the detection of intruders or hackers into a single computer or into a network; the creation of algorithms for detecting unauthorized changes in a system; and the intentional introduction of diversity into a system as a means for decreasing its vulnerability.

To design a digital immune system that incorporates all or most of these features, Forrest's approach requires the system to have an explicit and stable definition of self versus nonself. Further, the protection system should be able to prevent, detect, and eliminate harmful intrusions, such as computer viruses or hackers, remember previous "infections," and also recognize new ones. Finally, Forrest requires her computer immune system to have a means for protecting itself against self-attack (i.e., the digital equivalent of a lymphocyte attacking a self cell).

Recognizing Self and Nonself in Computers

The biological system uses several different mechanisms to differentiate its own cells (self) from harmful intruders (nonself). One way is mediated by protein fragments called *peptides* and is geared toward eliminating viruses and other intracellular infections; another means functions via white blood cells or lymphocytes, and attacks extracellular pathogens such as bacteria or parasites. In computers, given the range and diversity of possible security problems, it would make sense for a digital system likewise to incorporate several methods for distinguishing between self and nonself entities.

A digital definition of self would further have to be able to distinguish between legitimate changes to the system—such as adding or removing programs or users, or editing files—and unauthorized ones. Forrest's

computer immune system so far hasn't been able to perfectly differentiate between changes that are authorized and those that aren't, but she
believes that with a truly multilayered protective system, any unauthorized changes that slipped though would be caught in a subsequent layer
of protection.

Catching Unwanted Intruders

Forrest's approach for recognizing hackers and other unwanted intruders into the system can be applied not only to stand-alone machines, but
to a series of networked computers. She has chosen to apply the self
versus nonself paradigm to the problem of program security, rather than
gearing it to identifying authorized versus unauthorized users. In other
words, her methodology concentrates on achieving security in the system
program content—characterizing as self those programs already in the
system, along with any legitimate changes to them, such as adding more
memory or updating a piece of software, whereas nonself consists of all
other changes. It is these unwanted modifications to stored programs that
most likely indicate the presence of a virus or attack.

Just as the human immune system makes use of protein fragments
called peptides as one recognition mechanism for self and nonself,
Forrest's digital equivalent of peptides consist of short sequences of
signals between a program and the computer's operating system. By building up a database of normal signals for each program, it can monitor the
program's behavior. Each computer would have as many different databases that define *self* as it does stored programs, just as the body's *self*
consists of many different organs, tissues, and cells. In addition to using
system call patterns to represent self, Forrest and colleagues have also
experimented with other mechanisms including basic program code.

Implementing these ideas, she says, is time-consuming but not
intellectually daunting. For a given machine, each program's database
would show a pattern of usage that would constitute normal behavior
for that program. The presence of any abnormal usage patterns could
signal an anomaly. For example, if a program starts making inordinately
high demands on the operating system, it may be a sign of a virus or a
hacker.

Forrest calls her method fairly straightforward, and potentially very efficient. The immunized computer would use its resources primarily for checking programs and files that it is currently using—the most executed programs—without wasting time checking and re-checking those it hardly uses.

Distributed Change Detection

Forrest's second protection method mimics another aspect of the biological immune system, a key class of lymphocytes called "T cells." Detection of an invader occurs when molecular bonds are formed between it and receptors on the surface of the T cell. A range of different receptor types is needed to be able to recognize (or bind with) all the different kinds of pathogens that could enter the body. These various receptor types are generated at random, such that, in principle, there would be just as many self-recognizing cells as nonself-recognizing ones. In order to eliminate the former—and avoid autoimmune diseases—the body destroys them at an early stage of their lives before they've had a chance to mature and circulate.

By analogy, Forrest has designed a change-detection algorithm based on a similar principle of clonal selection. Her algorithm defines the notion of self based on a set of digital data, which can be monitored for any changes. In order to do this, the computer immune system would first generate a set of detectors that fail to identify self, and only recognize nonself. It would then use these detectors to monitor the protected data—a "hit" would mean that a change in the data must have occurred, which can be flagged and its location identified.

The detector's recognition method makes use of string matching between pairs of data strings. A perfect match between all the symbols in two data strings of equal length would constitute self. In reality, perfect matches would be rare, says Forrest, so the detector in effect would look for a certain number of contiguous matches between symbols in corresponding positions. Therefore, the total number of detectors needed to detect nonself would be the same order of magnitude as the size of self. Although effective, this method may not be practical for use in commercial anti-virus software as it requires too much storage space in the computer. Nonetheless, the algorithm can be distributed throughout a given machine, because, as a single detector would only cover a small segment of nonself, a series of them distributed throughout the system would produce a much wider range of coverage.

Either of these two methods, claims Forrest, could be part of a multi-layered system, located, for example, between a cryptographic system and user-authentication system for greater protection. They could also be distributed among several computers on a network. However, she cautions that these two methods embody only a few characteristics of the biological immune system, and feels more sophisticated properties will also have to be added for greater effectiveness. For example, immune cells are constantly replicating themselves; they are also generating new types, which would have to be mimicked in a computer immune version. In addition, the circulation pathways for immune cells in the body guarantee that any part of the body that gets injured will be repaired, so an appropriate analogue would have to be found for the computer.

Forrest disciple Steve Hofmeyr has started a computer security company, Sana Security, in San Mateo, California, which employs methods based on immune models. Sana has launched a software security product, First Response, which is geared toward protecting network servers and the information they store. First Reponse builds dynamic profiles of normal or routine behavior of the computer system (the self) and then detects anomalies or deviations from this behavior (nonself behavior) that likely signal an attack by an intruder.

Virus Hunting

Jeff Kephart and colleagues at IBM's Thomas J. Watson Research Center in Yorktown Heights, New York, began developing their vision of a computer immune system in the early 1990s. Ten years later, aspects of their digital immune system have been commercialized in the form of antivirus software, but according to Kephart, this is only the beginning of an effort to create a globalized computer immune system that can detect and eliminate a virus before it has time to spread. His bio-inspired security system mimics many of the same biological immune properties as Forrest's, in that it should be distributable, adaptable, multilayered, and incorporate memory as well as self versus nonself concepts. However, the two researchers' approaches show marked differences in the way they conceptualize the recognition problem.

According to Kephart, because the self recognition process in both human and digital immune systems is never perfect, both systems run the risk of mistakenly eliminating cells or programs that are not harmful. Because the body has so many cells and is constantly reproducing others, losing one or two self cells to an antibody probably doesn't matter. However, in computers, destroying legitimate software can be very costly. For that reason, Kephart feels that defining self solely in terms of software already on the system is not the answer. "Using a distinction of harmful vs. nonharmful, rather than self vs. nonself," claims Kephart, "is much more challenging. Any serious, commercially viable antivirus software must be very careful and clever in its handling of this issue."[1]

His differentiation process—determining what is benign and what is harmful—has several stages. First, an integrity monitor checks and flags any discrepancies between the original and current versions of the computer's data files and programs. An activity monitor then reviews the flagged items for behavior that is typical of viruses. Integrity monitors are also run constantly on the system to look for any indications of a virus. If they do spot something, it is checked against a list of known viruses.

This means their immune-like software would have to review the entire known corpus of computer viruses beforehand, extracting some sort of identifier from each one to compose the list.

If a virus detected is already known, it gets eliminated by proven methods. This aspect of the progam, says Kephart, mimics the function of the body's innate immune system, whose purpose is to respond to pathogens the body already knows. Its coverage is broader, but cruder and less specific than the adaptive immune system; it also takes longer to develop.

If the virus is not in this class, either the hit is a false alarm or else it represents a completely new viral infection. In the latter case, the bug is lured into a decoy program to see if it can infect it. If so, the digital immune system will attempt to extract an identifier or fingerprint for the unknown virus so it can check for its presence on other computers in the network. Ultimately, the new virus and its cure are sent to the virus database and added to the list of known infections, so that subsequent

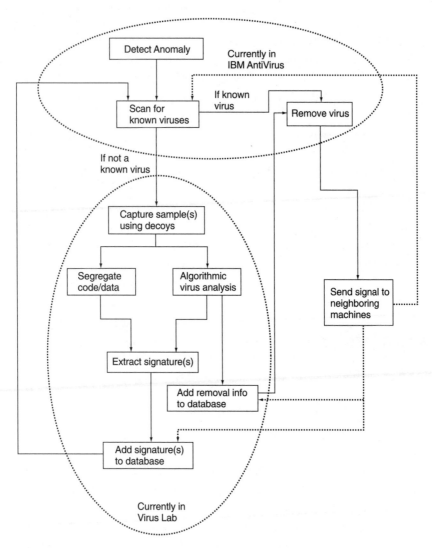

Figure 8.3
The main components of Kephart's computer immune system at IBM. (Image courtesy of Jeffrey Kephart)

uninfected computers can then be "vaccinated" against it. "This process is highly analogous to what happens in real biological systems," says Kephart, "in the sense that antibodies to newly encountered computer viruses are generated by molding themselves to the virus itself"[2] (figure 8.3). Kephart, who now leads IBM's antivirus unit, plans on continuing his work in computer immunology with IBM colleagues by transposing other aspects of the human immune system into the information domain. Together with Forrest, he feels there are many additional facets of the biological immune response that can be carried over to the digital domain.

9

Biologically Inspired Hardware

Speak to the earth, and it shall teach thee.
—Job 12:8

Researchers in the fields of computer hardware and electronics have also being drawn toward biology, where they have discerned an array of properties they can co-opt for their work. Bio-inspired hardware is defined as hardware that looks to biology for materials—either natural or modified—to serve as components in information processing or storage devices; or to inspire the design of computer components or architectures, even when made from conventional materials. The hardware included in the former category may be used in conjunction with traditional electronics, or even in place of them.

Though still at a fledgling stage of development, bioelectronics have the potential to provide significant advantages over traditional devices made with silicon. Design and construction of bioelectronics via genetic engineering or chemical synthesis can afford a much greater degree of control. Bioelectronic devices can also reach dimensions over one hundred times smaller than conventional silicon devices, down to the nanoscale (about one-billionth of a meter). Biology can also enable light-induced reactions at speeds two to three orders of magnitude faster than electrical switching in manmade materials, such as semiconductors. For example, in the protein bacteriorhodopsin, used for thin films, one photochemical process happens at 1 trillionth of a second. This capability could be harnessed for faster computer logic gates or light-activated switches. To further enhance their ability to serve as "intelligent" materials, biomaterials used in electronics could be genetically modified to customize their properties for a given function. There are still formidable challenges, however. Biomaterials have long been considered to be too unstable and fragile for use in electronics. Likewise, devising a robust method for interfacing a biomaterial with an engineered or artificial material (e.g., silicon, metal, glass, or ceramic) has also presented difficulties for researchers.

A Protein for Data Storage

Bob Birge, University of Connecticut biophysicist and director of the W. M. Keck Center for Molecular Electronics at Syracuse University, sees his work as part of the current trend for increasing miniaturization of computer circuitry, along with MEMS and nanotechnology. He has developed an ingenious yet complex protein-based data storage device that is now well into the prototype stage. Birge has strong convictions regarding the future importance of "biomolecular electronics," which he defines as "the encoding, manipulation, and retrieval of information at the molecular or macromolecular level."[1] He believes that proteins hold great promise as an intelligent material, with potential applications for future high-speed signal processing and communications, data storage, and artificial retinas, among other uses. Claims Birge,

Biomolecular electronics (bioelectronics) is a subfield of molecular electronics that investigates the use of native as well as modified biological molecules (chromophores, proteins, etc.) in place of the organic molecules synthesized in the laboratory. Because natural selection processes have often solved problems of a similar nature to those that must be solved in harnessing organic compounds, and because self-assembly and genetic engineering provide sophisticated control and manipulation of large molecules, biomolecular electronics has shown considerable promise.[2]

Birge chose this particular protein, bacteriorhodopsin, or bR, as a data storage material because of its unique photonic properties. (Photonics is an engineering field that deals with light-enabled processes, such as fiber optics.) Bacteriorhodopsin is also the simplest known biological pump, using light activation to pump protons across the cell membrane and thus produce energy. Bacteriorhodopsin is a purple pigment found in the membrane of a bacterium called *Halobacterium salinarium*, which inhabits salt marshes. Living in such a hot, swampy environment, this bacterium must function under high temperatures, high light intensity, and changing acidity levels for long periods, so it is robust and very efficient at creating energy via photosynthesis when oxygen levels are too low to permit respiration. When activated by light, the resulting cycle of chemical reactions serves remarkably well for data storage purposes. The total bR photo cycle takes about one-thousandth of a second, with some individual chemical sub-events even taking place a billion times faster that that. In addition, bacteriorhodopsin is both stable and very versatile.

Knowledge of bacteriorhodopsin and its suitability for electronic device applications is not new. In the Cold War era, scientists in the former Soviet Union discovered that very stable holographic imaging films could be made from bacteriorhodopsin, and some even felt that it might be one area where the Soviets could achieve technological superiority over the United States.

In Birge's view, technological advances in computing have caused the overall information processing bottleneck to lie more with memory devices than with processor hardware, providing an impetus for exploring new memory devices with better and cheaper storage capacity and higher bandwidths. Volumetric memory storage is a good candidates for this next generation technology, he believes, as it offers anywhere from 300 to 1,000 times the data storage capacity of traditional devices. Typical optical holographic memory devices, which make use of photosensitive crystals for data storage, are probably the most common volumetric storage device today.

Volumetric Memory Devices with Bacteriorhodopsin

Birge and his group have developed two different types of memory devices that use bacteriorhodopsin: a holographic associative memory, and a three-dimensional optical memory. The former utilizes the holographic properties of a bacteriorhodopsin thin film and is called an "associative memory" because it takes an input image and scans the entire memory bank previously stored in the thin film (called the "reference data") for the block of data that matches the input image (hence "associative"), similar to the way our brain works.

The 3D optical memory, which uses another aspect of photosynthesis, can hold data for a much longer time (about nine years or more), thus making optical data storage quite stable. Both systems store data at about 10,000 molecules per bit, with the speed of storage limited by the lasers needed for addressing the data. According to Birge, he and a graduate student discovered the bR-based three-dimensional optical memory quite by serendipity while in the lab.

Bacteriorhodopsin-Based Associative Holographic Memory

The three-dimensional associative holographic memory utilizes principles different than those used in traditional computer memories. Our desktop PCs have memories from which the programs and the data can be directly retrieved, and to which the resulting data is written, all in a sequential process. Associative memories work differently, more the way our own brains do. When the brain receives an input image from the visual system, it scans its entire memory contents simultaneously, and tries to associate the input image with a stored image, in order to recognize it. This process is very fast and very efficient. Holographic memory mimics the brain through a mathematical process known as Fourier transform association.

Before explaining this principle, it might be helpful to describe how a hologram stores data (figure 9.1). A laser beam, usually from a blue-green laser, is split into two beams: an object beam and a reference beam. The object beam, carrying the data to be stored, passes through a spacial light modulator (SLM), an LCD panel that converts the data into an array of dark and light pixels. A hologram is formed when the object beam meets the reference beam inside the photosensitive thin film or crystal. The storage medium captures the interference pattern from the two beams as a pattern of varying indices of refraction. Each page of data is recorded separately in this manner.

To read or retrieve a stored image, the reference beam is again formed and shone onto the hologram at exactly the same angle at which it originally entered the material to store the data. When passing through the stored interference pattern, it recreates the image on the original page. The reconstructed image is then projected onto an electro-optical detector for read-out. The key to this type of data storage is the accuracy with which the read reference beam matches the original reference beam that first stored the data. A bR-based thin film is used to store the data. Birge's prototype shows the entire apparatus for data storage (figure 9.2).

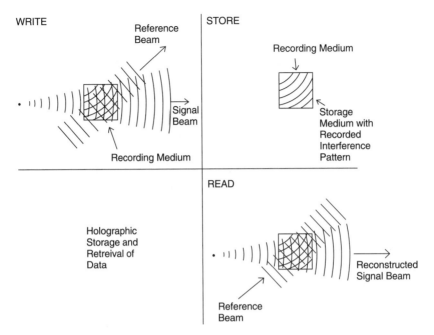

Figure 9.1
Holographic storage and retrieval of data. The "write" phase: the object beam coming from a point-like source meets the reference beam in the storage medium and forms an interference pattern. The "store" phase: this light pattern modulates locally the refractive index of the recording medium, resulting in stored data. The "read" phase: to retrieve the stored information, the medium is illuminated by the same reference beam. Due to refraction by the stored refractive index grating, the original object beam is reconstructed.

In this system, the associative principle forms the basis for the image retrieval. The input image is compared optically to the block of images stored in the hologram. The input image will share a number of common characteristics with its closest match, enabling (through constructive and destructive interference processes) the closest match to be preferentially selected over other holograms stored in the database. This preferential selection is manifested by an increased intensity in the light transmitted through the location in the holographic database where the most closely matching image is stored.

Real-world applications of this type of associative memory include fingerprint identification and counterfeit currency detection, as well as the ultimate goal of true artificial intelligence. It is generally recognized that

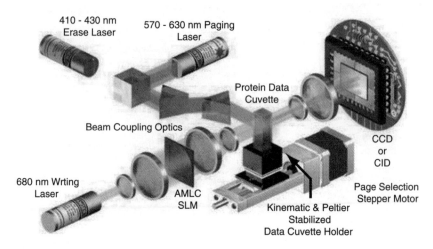

Figure 9.2
Bob Birge and his group at the W. M. Keck Center for Molecular Electronics at Syracuse University have developed an optical holographic associative memory storage device that makes use of a thin film made from a protein, bacteriorhodopsin. This device is called an associative memory because it takes an input image and scans the entire memory previously stored on the thin film (called the reference data) for the block of data that matches the input image. The reference data is stored as enhanced Fourier transforms in the protein-based thin film.

artificial intelligence will require large-scale associative processing with enormous databases, if it attempts in any way to mimic the neural associative memory of the human brain.

The advantages that bacteriorhodopsin provides as a holographic medium are sensitivity and longevity. Bacteriorhodopsin can respond quickly to slight shifts in the intensity and phase of light, and immediately register them in the resulting hologram, which is rare for holographic data storage materials. Longevity refers to the number of times the material can be used (i.e., written, read, and erased) before loosing its effectiveness—in bR-based holographic storage, the film can be used almost a million times before losing its abilities. It also has very high resolution for images, with more than 2,000 lines/mm.

Bacteriorhodopsin and Optical Memory

The second method, 3D optical memory, relies on the sort of serial memory used in computers rather than on associative principles, and can store 300–1,000 times as much data as other memory devices of the same size. Birge's bR-based version, as all random access memories, can read, write, and also erase data.

The method used for data storage is quite ingenious. The storage medium is a transparent container about 3 cm³ in volume, containing a thin film made from bacteriorhodopsin. Laser light is shone onto it, activating a very thin 2D region of the film. Each of these regions will constitute a "page" of data that can store over 16 million bits of data. The laser pumps light energy into bR molecules in that page, exciting it to a higher, yet not very stable energy state, called the O excited state. Two milliseconds after the initial laser pulse, another laser with a different wavelength, located at right angles to the first, is switched on and irradiates certain sections of the same page that will represent the one's in the binary pattern of the input data. The molecules in those spots are also pumped up, but this time to a more stable excited state, the Q state. These molecules remain in this state for a long time, while the remainder of the molecules on that page, which are in the O state, lose their energy and return to their original resting state. The resting state then represents the binary "zero," whereas the Q state represents the binary "one" state. (The initial O state was activated to widen the energy gap between the Q state [the ones] and the ground state [the zeros], to which those molecules in the O state quickly revert.) This entire phase constitutes the data storage phase.

To read the data, the reference beam is again fired into the container; after two milliseconds, the object laser is turned on, but this time at low intensity. (This prevents the molecules from being pumped up into a Q state.) Molecules representing zeros absorb this light, whereas those representing ones don't. The light and dark absorption pattern of the page is imaged onto a detector, which captures it as a page of digital information. The data can be erased and reused a little over a thousand times.

Birge estimates that data recorded in this way on bacteriorhodopsin thin films can last about nine years. His group has produced three prototypes of the 3D memory, and hope to improve the storage density by fivefold. Although the memory is still too inefficient, they have studied over six hundred mutated forms of bR during the past two years, and are confident that it will be a commercially viable system in a few years.

Both NASA Ames Laboratory and Motorola are also experimenting with bacteriorhodopsin thin films for various purposes, including data storage. Eager to get a jump on any potential competitors, a German company, Consortium fur elektrochemische, Industrie GmbH, has even put bacteriorhodopsin on the market for what they advertise as "optical data processing, holography, light sensors," and other applications.

The value added in protein-based memory, says Birge, is that bR is cheap to produce, and genetic engineering can boost its performance. Furthermore, bR-based memory is extremely rugged and can withstand electromagnetic radiation at far higher levels than semiconductor memories. The data storage is very stable and retains all its information even when the system is shut down—so it could make computers very energy efficient. The data cubes are small and easy to transport, without the kind of fragile, moving parts in a hard drive or cartridge. With bR-based memories, huge amounts of data could easily be carried from one place to another. Birge feels the outlook is promising for protein-based data storage, given its advantages, and will increasingly improve as familiarity with bR grows.

Harnessing the Sun's Rays in Arizona

Researchers at Arizona State University are not only developing their own version of biomolecular hardware, but have also established a multidisciplinary program of research and graduate study in this field, with the help of the National Science Foundation. The program integrates biology, biophysics, chemistry, and engineering, with a focus on biomolecular devices, both natural and manmade, and offers training in biohybrid circuits, light-powered molecular engines, DNA synthesis and repair, and protein engineering. Arizona State University (ASU) has been a well-known center of research activity in photosynthesis for over fifteen years, with a large group of faculty from the life and physical sciences engaged in this work. "We came to the point," says ASU scientist Neil Woodbury, "where it was clear to us that the paradigms we were learning in the study of photosynthesis were suited to a large number of electronic device applications. We wanted to work on these problems, so starting a program with students, classes, and seminars, seemed like a good vehicle. Furthermore, the idea captured our imagination."[3]

Photosynthesis—the process that occurs in green plants when it synthesizes organic compounds from carbon dioxide and water in its environment in the presence of light—can be viewed as a light-activated power source in nature. It involves several processes that can be co-opted for electronics. For example, it happens very fast (roughly a millionth of a millionth of a second, compared with computer clock times that are in the order of a billionth of a second), and is switched on by light, suggesting the possibility of photosynthesis-powered optical logic gates—a true organic computer!

Arizona State University researchers Devens Gust, Tom and Ana Moore, and Gali Steingberg-Yfrach have studied the activities of the photosynthetic "reaction center," a center where photosynthetic light produces energy. They realized that the process could be a microscopic solar power pack, and so set about trying to create a synthetic version. Their work employs water suspensions of liposomes, which are tiny hollow spheres with double-layered lipid walls, resembling cell membranes. The group has learned how to add synthetic compounds to these walls, causing them to act just like the photosynthetic membranes inside plant cells. The liposomes are then able to capture sunlight electrochemically, using it to generate energy. "These artificial systems," says Gust, "may be useful for powering small man-made machines that couple light energy to biosynthesis, transport, signaling, and mechanical processes."[4]

However, once they've made these photovoltaic devices, researchers must still find ways of interfacing them with electronic circuits. There are several processes that could serve as candidates for this task: absorbing or emitting photons, heat intake or loss, or electrical signals from the movement of electrons. So far, however, they have found no reliable form for the interface, and, says Neil Woodbury, "No one says it's going to be easy."

Another effort at ASU's biomolecular center is more directly related to transistor applications. Working with Michael Kozicki, a professor in the Electrical Engineering Department, Gust has developed a hybrid device that uses surface monolayers of light-activated molecules to control the current flow in a transistor. Kozicki strongly believes that the first practical applications of these molecular electronic systems will require a novel architecture different from that used in current silicon transistors. Closer scrutiny of biological systems will help guide these efforts, says Kozicki. "I'm the first to admit that biology is the best teacher here."[5]

Biology on a Chip

"Nature is the grandest engineer known to man—though a blind one at that: her designs come into existence through the slow process, over millions of years, known as evolution by natural selection."[6] Moshe Sipper speaks with a respect that borders on awe when referring to nature's ability to create living systems that can produce energy, process information, self-replicate, self-repair, function with defects, adapt to their surroundings, and improve the species over time through evolution.

Sipper, colleagues at the Swiss Federal Institute of Technology (EPFL) in Lausanne, Switzerland, and several other groups of researchers around the globe have undertaken a purposeful study of these natural processes in order to try and translate their underlying principles into the language of information technology—specifically, those aspects that might have a bearing on integrated circuit design. "Nature has found effective ways to deal with large, complex systems, the numerous elements of which function imperfectly. That's where artifacts are going now, specifically electronic hardware," muses Sipper.[7] Three types of bio-inspired methods for the design of novel computer chips are evolvable hardware, embryonic hardware, and "immunotronics," each modeled on different properties of biological systems.

Evolvable Hardware

In addition to the group at EPFL, researchers such as Adrian Stoica of NASA's Jet Propulsion Laboratory (JPL), have also been pursuing a vision of a computer chip as an adaptable, biological organism that can be programmed to evolve to an optimum configuration. This approach, dubbed evolvable hardware (EHW), is defined as hardware whose structure, function, and architecture change over time, in an autonomous manner, in order to optimize its performance.[8]

Evolvable hardware does not mean the use software with evolutionary algorithms to simulate the iterative improvement of circuit designs.

Rather, it entails changes in the actual structure of the hardware, online or in a test environment, due to its ability to reconfigure its structure dynamically and autonomously. The progress of EHW is inextricably linked to advances in the fields of evolutionary computation and reconfigurable circuits. Whereas with conventional hardware, the components, circuitry, and interconnects on the integrated circuit are fixed and unchanging, EHW attempts to adapt to the requirements of real world problems that can change over time—and thus maximize its performance for a desired task. Although programmable logic devices already exist that are being used by chip designers in early design and development stages, these chips require precise hardware specifications a priori based on the requirements of the operating environment. With EHWs, on the other hand, one can design chips with the desired functionality, without having to take into account the changing operating environment, because they can change their own circuitry via evolutionary computation (figure 9.3) according to changes in task requirements or in the operating environment. The ultimate application of an evolvable chip, say its most

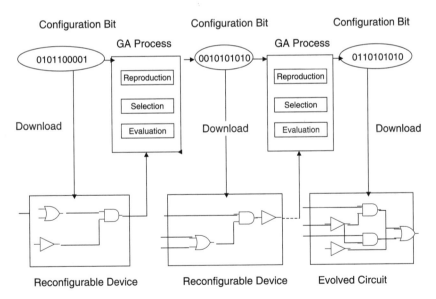

Figure 9.3
The basic concept of evolvable hardware is based on the idea of using reconfiguration bit strings as the "DNA" for running genetic algorithms, which makes use of a fitness function to evolve the optimum integrated circuit configuration.

optimistic proponents, will be its use in the creation of the artificial brain.

Whether the EHW concept has long-term commercial potential or is ultimately destined to die off as an extinct branch of computer science still remains to be seen, although early results look promising. Though some computer scientists have voiced concern over the black box nature of EHW—the difficulty in analyzing, understanding, and reproducing the circuits evolved—researchers claim they are currently addressing this problem, and that some amount of skepticism is normal for any nascent technology.[9] NASA seems ready to bet on the ultimate success of EHW, and has already held several conferences on the topic. Their interest is due to the autonomy of EHW in reconfiguring its circuits to produce the optimum configuration, a property that bodes well for applications in space where humans can't intervene.

"Evolvable hardware is an extremely exciting new field," says Jose Munoz, former DARPA program manager, and now director of the Simulation and Computer Science Office in the Department of Energy's Defense Programs Office.

However, it should be approached with a caution that hopefully won't temper the excitement. There have been some results reported which couldn't be duplicated because the effects were the result of some aberration in the materials used. It is imperative that the process, techniques, and approaches used be reproducible if they are to be of any general value, and not just, "Isn't this an interesting effect!" The danger is that EHWs often come across as ad-hoc.[10]

Why Use EHW?

Computer manufacturers are always forced to make trade-offs between performance and the cost of manufacturing. Although a highly specialized computer can be very efficient and extremely fast at the specific function for which it was designed, building a one-of-a-kind, custom machine can be very expensive and time-consuming. So computer companies today tend to settle for general-purpose systems, which, although they are cheaper and more practical, must ultimately sacrifice something in performance capability. And though the chips an all-purpose computer uses are quite versatile, their hardware remains fixed. Evolvable hardware, on the other hand, has circuitry that can be reconfigured and improved during operation, and thus for a given cost, can potentially eke out a higher level of performance.

Another advantage of EHWs over existing methods of circuit design based on either human workmanship or circuit design algorithms is that when a circuit is designed using simulations, it doesn't always work exactly as predicted when finally built. Evolvable hardware designs, on the other hand, actually emerge on the chip itself. And although EHWs still rely heavily on software to control the evolutionary process, some scientists envision the day when everything could be put onto the chip, including the genetic algorithm, which would make the whole process exponentially faster. Putting thousands of chips like this together could even simulate a learning process like the brain's.

How EHW Works

Generally speaking, most EHWs rely on a class of computer chips called field-programmable gate arrays (FPGAs) in order to work. These are arrays of configurable digital or analog chips, which are very fast-acting and whose circuit designs can be modified or configured using a special set of software tools (figure 9.4). The software reconfigures and routes the logic on the FPGA at three different levels: in the function of the logic cells themselves, the interconnections between the cells, and the inputs and outputs. The bits themselves can be reconfigured in nanoseconds, and the entire chip reconfigured online. The secret to performance in these devices lies in the logic contained in the logic blocks and in the performance and efficiency of their routing architecture. The first FPGAs were invented by Xilinx Corp of San Jose, California, in 1984.

A "reconfigurable" circuit is based on the idea of a string of bits representing a given configuration. This is fetched from the computer's memory and is directly used to configure the hardware, with no need for further elaboration. Some configuration strings do not change during the execution of the task, whereas others, termed "dynamic," can change in operation, as the circumstances change. Evolvable hardware uses the latter type of bit string.

There are two ways to evolve an optimal circuit design. One is restricting the evolutionary process to designs that resemble conventional designs, and are hence easier to analyze. The other is giving the

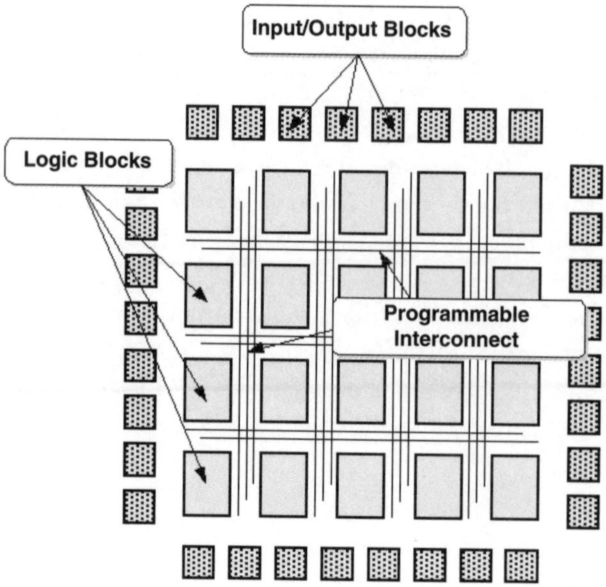

Figure 9.4
Field-programmable gate arrays (FPGAs). FPGAs are digital integrated circuits whose circuit design can be modified or configured using a special set of software tools. They consist of an array of logic blocks that are connected together via programmable interconnects, and blocks for input and output. (Source: Xilinx, Inc.)

evolutionary process free reign to explore any design space it wishes, even the most unconventional and bizarre. Whereas in the first instance, the search-space is constrained to circuits whose designs are fairly conventional, albeit understandable, in the second, they can virtually be anything they want. The latter approach is somewhat risky, but may result in better solutions.

As an example of the first approach, imagine a chip that you want to be able to recognize the numbers 1–9. You would start with bits strings that would specify perhaps five hundred different circuit designs, and the fitness criteria of the genetic algorithm would tell you how well each one recognized a given number. The genetic algorithm would download each configuration onto the FPGA, test the resulting circuit design, and after testing them all, keep the best ones, which would be mated together to produce the optimal offspring of a number-recognizing computer chip.

As for the second approach, a group of computer scientists at Britain's University of Sussex, led by Adrian Thompson, have taken it to an extreme. Their design space is not only completely unrestricted, but instead of keeping things solely in the realm of digital logic gates, their circuits can adopt a range of intermediate values between the standard binary 1 and 0. In other words, their EHW doesn't direct the path of evolution at all—it just lets the process itself mysteriously come up with the best circuit design it can. Although very adaptable, the problem with this approach is that the process can become *too* mysterious and it often becomes extremely difficult to try and reverse engineer the chip to understand how it works. If the chip malfunctioned in a critical application such as in a medical device, for example, understanding what went wrong would be crucial.

Applications of EHW

Several applications of EHW have been implemented so far with some degree of success. One is in communications, certain areas of which still require high-speed analog circuits.[11] One problem with analog circuits is that the values of some of the final manufactured circuit components, such as resistors and capacitors, often differ from initial design specifications. For example, in intermediate frequency filters, which are often used in cellular phones, the tolerance level is less than 1 percent. This means that any analog circuits that don't meet this requirement must be discarded. However, with an analog EHW chip, the variations in the circuit values can be corrected during the test phase, or even after fabrication, thus increasing the yield rate. Another advantage of using EHW here is that the analog chip components can be smaller, resulting in lower manufacturing costs and lower power consumption. The smaller size is also obviously well suited for cell phone use.

Another use for EHW is in electrophotographic printing, a technology that makes it possible to print books with high-precision photo quality.[12] Data compression is essential here. For example, one 1,200-dpi electrophotographic image requires 70 MB for storage, so if an EP printer processes a hundred different pages in a minute—a standard speed for this machine, which is ten times faster than a regular color printer—7 GB of data must be transferred to the printer. This must happen at a speed of 1,800 MB/minute, or about six times as fast as a typical hard drive.[13] These printers therefore have to be very efficient at data compression, and also be able to reconstruct the compressed data into an image very quickly. Most traditional data compression methods aren't up to the task.

In Japan, a group headed by Tetsuya Higuchi of the governmental Electrotechnical Laboratory and Nobuki Kajihara of NEC Laboratory have collaborated in producing an EHW chip for data compression that changes entire cells at each step, rather than individual bit strings (the DNA of evolutionary algorithms). They use an FPGA whose bit strings, rather than resetting individual logic gates, change entire functions, such as adding or subtracting. Higuchi believes that this method produces chips that are more useful for applications. Their EHW for data compression has achieved compression rates twice that of international standards by using a precise prediction mechanism to reconfigure the circuits. Higuchi has also been successful applying EHW methods to correct faulty circuits in analog chips, thus increasing the yield.

Erratic Behavior

There are downsides, however, in addition to the fact that with an unconstrained search space, the evolutionary algorithm creating the final circuit design can be so complex that it becomes hard to analyze and replicate— a real problem when the chip malfunctions and has to undergo troubleshooting. Some users also complain about the intermittent behavior of EHW over long periods. The chip must be able to function uniformly over time, and be insensitive to changes in environment, or else be able to adapt to them. However, conditions such as thermal drift, noise, and aging effects in semiconductors devices can sometimes cause erratic behavior in EHW chips.

Says Caltech's Andre DeHon, an expert in circuit design, "An evolved chip may work fine at one temperature, but break down as the temperature changes, as for example, in its timing mechanisms. This doesn't happen with conventional hardware, where a careful set of disciplines and models are used to make sure the operation is robust to environmental changes."[14] Advocates contend, however, that EHW can adapt to temperature changes by evolving a new design better suited to the new operating environment.

Adrian Thompson and Paul Layzell at the University of Sussex have come up with a toolbox of methods that will enable the user to cobble together an analysis of even the most evolved or bizarre EHW chip. This includes such means as testing the chip under abnormal conditions, looking for parts of the chip design that are amenable to mathematical analysis techniques, simulation of the circuit, monitoring power consumption or electromagnetic emissions from rapidly changing electrical signals, and studying its evolutionary history or the similarities in a given evolutionary generation of circuit designs.

With caveats such as these, however, does it even pay to chose an EHW over a more tried and true alternative? Caltech's De Hon talks of having struggled with circuit design problems where there are many variables to consider, such as function, durability, and the cost of all resources needed. He feels the real benefit from EHW arises when the solution space and cost functions for a proposed circuit model are not well defined. "In such an undefined space as this, with unclear or changing cost functions—what I'd call an NP-hard problem—EHWs work well."[15]

JPL's Adrian Stoica has focused his EHW research efforts on applications in unmanned spacecraft, where chips must be able to autonomously self-repair or else adapt to changes in the operating environment. They have been able to show successful adaptation to higher temperatures and also very fast times (tens of seconds) for new circuit evolution.

Embryonic Hardware

Instead of Darwinian evolution, the operative biological metaphor here is embryogenesis, the process through which the fertilized egg eventually develops into a complete organism, by continually dividing, copying, and transferring the genome from the mother to the daughter cell with each division. Eventually, these cells will differentiate into distinct kinds of cells (there are about 350 different types of cells in a human), some becoming a liver cell, a muscle cell, a joint cell, or a skin cell, with different forms and functions—called morphogenesis.

Cellular differentiation depends on what part of the genome gets interpreted, which in turn is contingent upon its physical position within the cell. Each cell in an organism is "universal," containing the same genome, thus making it capable of self-repair and self-replication. Says Adrian Stoica, "As we move towards the idea of imperfect components in processing structures, the embryonics concept will find its way into many silicon and nonsilicon designs of tomorrow."[16]

Daniel Mange and colleagues at EPFL in Lausanne, including Moshe Sipper, have devised an electronic analogue of this process, called an embryonic circuit. They wanted to create a chip that could repair and even copy itself, which would be useful when highly robust, reliable chips are needed. Their model, embryonics, adopts certain features of cellular organization, transposing them to the two-dimensional world of integrated circuits (figure 9.5). For properties such as self-repair, this

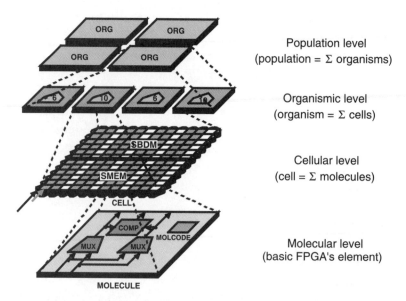

Figure 9.5
Hierarchy of "embryonic" elements. Embryonic circuits, in analogy to living organisms, are part of a hierarchy. The basic unit is the FPGA, which parallels the molecule in living organisms. Many FPGAs make up an embryonic cell, which in turn, in the aggregate, constitute an organism. Many embryonic "organisms" make up a population. (Image courtesy of Daniel Mange of the Swiss Federal Institute of Technology)

means defective elements can be partially reconstructed, whereas self-replication means it is possible to rebuild an entire faulty component. This can be very valuable in applications where reliability is critical, such as in avionics, medical electronics, at geographically remote locations, or in space where there are high levels of radiation. Embryonic circuits may also be useful for the future hardware of molecular electronics where the components cannot be fabricated 100 per cent fault-free. Declares computer scientist Andy Tyrell of Britain's University of York, who has collaborated with the Lausanne group, "Embryonic circuits would also work well in applications which exploit the latest technological advances, such as drastic device shrinking, lower power supply levels, and increasing operating speeds—all those things that accompany the evolution to deeper submicron levels and increase the error rates."[17]

The Biowatch

The EPFL prototype machine is "Biowatch," an artificial organism designed to keep time. It is one-dimensional and comprised of four identical cells. Just as in nature, each cell carries out a unique function, according to its genome. Each "gene" is actually a subprogram, characterized by a set of instructions and by a horizontal coordinate, X (see figure 9.6). Each cell contains an entire genome, just like its living

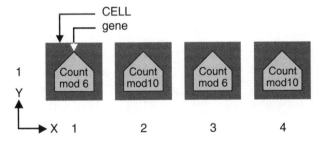

Figure 9.6
Multicellular organization of the "Biowatch"—an application of embryonic circuits for telling time, capable of self-repair and self-replication. It can count minutes (from 0 to 59) and seconds (from 0 to 59); it is thus a modulo-3600 counter. The Biowatch is multicellular, that is, it is composed of four cells with identical physical connections and containing identical "genes." Each gene is a subprogram, characterized by a set of instructions and by its horizontal coordinate, X. (Source: Daniel Mange)

counterpart, and depending on its position in the array, will carry out only a designated part of the genome.

Biowatch exhibits "cellular differentiation," in the sense that each cell can perform one of two tasks: a modulo-6 or modulo-10 count. A modulo operation in arithmetic is when the result is the remainder after one integer is divided by another. Hence i modulo j is the remainder of the division of integer i by integer j (see figure 9.7). When faults occur, self-repair is achieved by reconfiguring the task some of the cells execute. A KILL signal detects the faults and is sent out by the cell via a built-in self-diagnosis mechanism. When the fault is located, the entire column containing the fault is deactivated. The organism must have as many spare cells to the right of the array as there are injured cells in order to repair itself. The faulty cell is then bypassed, and the remaining parts of the original cell are shifted to the right (figure 9.8). The new configuration leads back to the original task (e.g., modulo-6 or modulo-10 count). By adding more cells, Biowatch can be used for practical time keeping, and doesn't miss a millisecond while self-repair is taking place.

The Lausanne group is now developing a new electronic tissue for a giant Biowatch. Says Mange, "This electronic array is a totally scalable assembly of molecules, each being a three-layer sandwich made up of a touch screen button for input, a display (output), and an FPGA. Applications will be intelligent/interactive walls, blackboards or slates, and later flexible tissues used for clothes."[18]

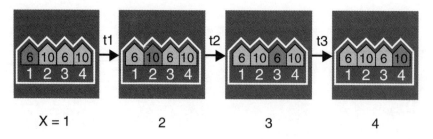

Figure 9.7
Cellular differentiation of the Biowatch. Though each cell contains the same "gene" or subprogram, each interprets a different part of it, depending on its position within the organism. (Source: Daniel Mange)

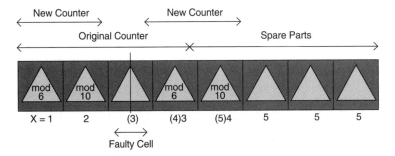

Figure 9.8
Self-repair of the Biowatch. The old coordinates are shown in parenthesis.
(Source: Daniel Mange)

An Immune System for Chips

"Immunotronics" is a new project that looks to design more fault-tolerant computer hardware based on the principles of the human immune system. Fault tolerance is a measure of how reliable a computer is and to what degree it can tolerate defects, faults, or malfunctions and still continue to function as it should. Entire computer systems can be designed to be fault tolerant, as well as their subcomponents, such as networks, software, or integrated circuits. A common method of building fault tolerance into a system is through redundancy, using two or three of the same systems, subsystems, or components placed in parallel. For example, in NASA's space shuttle, several layers of redundancy in their computers ensure that the likelihood of total system failure remains small. This sort of redundancy has its drawbacks, however, as when a single logic gate on a chip breaks down and an entire subsystem or module has to be replaced in order to get the system working again—an approach that can be both impractical and expensive.

Andy Tyrrell of UK's University of York chose to model his fault tolerance efforts on natural mechanisms. "I have always been impressed by the remarkable complexity and precision of the genetic process and by its robustness in the face of change, errors, and injuries," he muses. "In addition, our understanding of how biological organisms manage to survive both individually through self-repair and from one generation to the next through evolution has dramatically increased over the last few years, giving researchers new models for computing," says Tyrrell. "Tolerating faults and handling complexity are two of the main challenges of the design of next-generation electronic circuits."[19]

Tyrrell's immunotronic (electronics with immunological properties) information processing systems are highly distributed with a large degree of redundancy. Like the human immune system, they protect themselves against faults through an ability to differentiate between self (healthy cells and tissues, in the body's case) and nonself (dangerous intruders, such as viruses or bacteria). The concept of distinguishing between self and nonself has already inspired novel approaches to computer virus protection, network security, and software fault tolerance with varying degrees of success (see chapter 8).

According to JPL's Adrian Stoica, "This approach is still very new, and still in a definition phase with no hardware demonstrations yet." However, he believes it has potential in specific areas, especially when combined with embryonics. "This would take the form of an interactive network model in which each embryonic cell is monitored by several immune cells, which is currently under study. Evolvable hardware techniques may also play a role in shaping such hardware immune systems, especially in the training of the detection mechanisms."[20]

The immunotronic method involves three phases: detection of a fault or malfunction in the hardware, minimization or suppression of the effects of the fault, and activation of an appropriate strategy for "healing" the fault. In order to render the system fault tolerant, it must be put through a learning phase, where it is trained to recognize self versus nonself states. This phase usually takes place when the system is being tested and developed, prior to actual operation. The step parallels the maturation of a certain type of immune cell or lymphocyte in the body called a T cell. The data characterizing self and nonself states needs to be broad enough to cover a wide range of anticipated faulty states. Afterwards, when the system is actually running, states and their transitions are monitored by comparing them to data stored in the memory. A faulty state will be flagged and acted upon accordingly, as soon as the system recognizes it as such.

The complex part comes when the system encounters a fault that is not precisely like one stored in its memory. Exact fault matching is relatively easy to accomplish, albeit very memory intensive, but the system also needs a methodology for dealing with unmatched faults. There must also be a clearly defined margin of tolerance to determine which faults are negligible enough not to be tagged as bonafide faults (or recognized as self states) and which ones aren't. This capability for partial fault matching is one of the attributes that Tyrrell believes makes this method valuable, giving the system more flexibility and a capability to flag and process an extremely wide range of faults.

Tyrrell and others have implemented both exact and approximate matching systems, using for the latter a negative selection algorithm inspired by the work of Stephanie Forrest at the University of New Mexico and the Santa Fe Institute, and her colleagues (see chapter 8). Maintains Tyrrell, "In the end, living systems have had several billion years of evolution to optimize their own special version of information processing systems. Our work in bio-inspired computing attempts to 'steal' a small fraction of this work for the advancement of technology. In essence, we are studying the past in order to create the future."[21]

Whither Computer Hardware?

EPFL's Moshe Sipper and others involved in designing bio-inspired information processing systems hope to cull further inspiration from the book of nature in their future work on EHW, embryonic hardware, immunotronics, and other areas of reconfigurable hardware. Sipper himself predicts that the merger of classical processors and reconfigurable hardware will intensify, possibly to the point where the all the working parts of a computer can adapt to their environment (i.e., everything they "touch," including other components, the user, and the network) and even self-repair. There are of course obstacles ahead: configuring a processor is not easy, and the technology isn't mature yet—but proponents seem confident that the field will evolve and eventually surmount these problems. Reconfigurable hardware, in their view, will be what ultimately enables the future of information processing systems.

Neurally Inspired Hardware

Being the miscellany that it is, biohardware can comprise almost any kind of device whose structure or function resembles or incorporates a biological object or process. In this final section, we give a brief overview of several highly innovative and significant research efforts that look to the mammalian brain for their inspiration.

When describing the brain's mode of processing information in 1948, when little was known about the physiology of neural information processing, von Neumann correctly described its function as both digital and analog. The digital mode derives from the essentially on/off way a neuron fires, whereas the analog behavior is reflected in the fact that not only can the rate at which neurons fire vary continuously, but they can also exhibit an analog response to conditions such as temperature, fatigue, stress, and so on. Several researchers have developed silicon-based hardware that is closely modeled on these properties.

Caltech's Carver Mead, Kwabena Boahen of the University of Pennsylvania, and Chris Diorio of the University of Washington, the latter two former students of his, have been working in a field known as neuromorphic engineering, in an attempt to mimic the organizing principles of the brain. Says Boahen, "The computer uses a hundred thousand times more devices, and associated wires, for each instruction, than the brain does for each event. Hence, we can improve the energy efficiency of present-day computers dramatically by copying the organizing principles of the nervous system."[22]

Carver Mead, now emeritus at Caltech, and one of the inventors of VLSI systems (integrated circuit fabrication technology that allows over a hundred thousand transistors to be integrated on a single chip) has experimented with neuromorphic electronic systems, which are closely modeled on cerebral functions. He and his students hold key patents on systems modeled after human vision, hearing, and learning mechanisms. According to Mead, products based on these principles have the potential to transform the interface between computers, images, sounds, and the user. However, he feels progress in this direction in hardware development has been lacking, which is one of the reasons he shifted his research goals fifteen years ago from digital circuits to studying human neural function.

A electronic neuron chip, according to Mead, makes comparisons between stored images and input from the real world, then reports on its findings—a vastly more sophisticated mode of functioning when compared to today's computers, which are based on chips that can only essentially manipulate ones and zeros. As he explains, many neural processing areas are organized as thin sheets, which give two-dimensional representations of their computational space, so these structures map well onto a two-dimensional silicon surface. Furthermore, in both neural and silicon technologies, the active devices (i.e., synapses and transistors) occupy no more than one or two percent of the space, with the remaining space filled with "wire."

Former Mead disciple Boahen is currently working on developing a silicon chip that imitates the function of the retina by converting photons into spiking neurons that encode spatiotemporal messages. The end goal would be to create a high-fidelity, single-chip visual prosthesis. The size, weight, and power consumption of the chip must be comparable to that of the retina. Boahen's chips recreate the structure and function of neural systems in silicon using standard VLSI and CMOS technology, starting from the device level through the circuit level to the system level. At the device level, the analogue for electrodiffusion of ions through membrane channels is electrodiffusion of electrons through transistor channels. At the circuit level, Boahen has developed an implementation mimicking synaptic behavior, integration of the dendritic signals, and behavior that reflects active membranes; and at the system level, he is able to synthesize the spatiotemporal dynamics of the cochlea, the retina, and early stages of cortical processing.

University of Washington's Chris Diorio, also trained by Mead, continues to work on his goal of creating electronic systems that employ the computational and organizational principles used in living nervous systems, which he calls "silicon neuroscience." He and his colleagues have started modeling the ability of the brain to adapt and learn, including its mechanisms for long-term memory, synaptic plasticity, and neuronal growth. The group has developed a type of CMOS, called the synapse transistor. These devices function like neural synapses, and can perform long-term nonvolatile analog memory, allow bidirectional memory updates, and can learn from an input signal without interrupting the ongoing computation. These synapse transistors also compute the product of their stored analog memory and the applied input. Although Diorio doesn't believe that a single device can completely model the complex behavior of a neural synapse, he feels their synapse transistor can implement a local learning function.

William Ditto, a physicist at Georgia Tech, agrees that living neurons do a much better job of processing information than a computer does. However, instead of trying to create silicon devices modeled after neurons, he envisions a hybrid biocomputer that combines living nerve cells with VLSI. Ditto and his group have used large neurons from leeches in vitro, and connected them up to a PC by microelectrodes in order to send signals to each cell, representing simple numbers. In order to get their simple hybrid biocomputer to perform a basic arithmetic calculation, he used elements of chaos theory to stimulate the leech neurons. The neurons made connections among themselves to "compute" their own way to solving the problem. The PC was then able to extract a comprehensible (and correct) answer from the noisy traffic that followed. Ditto feels this approach may be particularly suited to pattern recognition tasks, such as reading individual handwriting.

Daniel Morse, chairman of the Marine Biotechnology Center at U. C. Santa Barbara, has long been studying the behavior of marine animals (e.g., mollusks with shells, pearls, corals, marine sponges, and diatoms), and realized that their ability to synthesize and secrete composite materials with extreme precision could be harnessed to develop new strategies for the creation of high-performance, nanostructured composite materials for use in advanced optoelectronics, microelectronics, catalysts, sensors, and energy transducers. By studying the molecular mechanisms of these marine organisms, in both calcium-based and silicon-based systems, researchers have been able to come to a greater understanding of the processes of biosynthesis and supramolecular self-assembly so as to apply them to the development of new biomaterials. For example, proteins purified from the exceptionally strong microlaminate of the pearly abalone shell have served to control the creation of a polymetallic, crystalline thin-film semiconductor. Morse and his group have also discovered that the glassy needles of silica made by a marine sponge can catalyze the synthesis of opal-like silica and silicone polymer networks that could potentially create high-performance silicon-based materials for computing.

10

Biology Through the Lens of Computer Science

We have come to the edge of a world of which we have no experience, and where all our preconceptions must be recast.
—D'Arcy Wentworth Thompson, *On Growth and Form, 1917*

The last few decades have witnessed a veritable flood of experimental data and new discoveries emerging from the life sciences. The genomes of over a hundred organisms, including the human genome, have now been mapped in their entirety. (The genome gives the full complement of an organism's genetic material or DNA. In humans, it is the source of our uniqueness and is contained in each of the twenty-three chromosomes we inherit from our parents.) Because the genome contains the code of life—that is, the instruction set that directs the manufacture of proteins needed for cellular growth and functioning—its decoding represents a major step toward unlocking the secret of life. Some say we are in the midst of a genomic revolution, equal in magnitude and impact to the information revolution.

New specialized fields of inquiry, such as genomics, the study of the genetic material needed to specify an organism, and proteomics, the investigation of entire protein complements in cells and organisms, have arisen that are aiding us to better understand how cellular phenomena arise from the connectivity of genes and proteins, and consequently how our own system functions—and malfunctions. Little by little, scientists are beginning to map some of the simpler, albeit still highly complicated biochemical signaling networks inside the cells that enable gene-gene, gene-protein, and protein-protein interactions. These rapidly occurring advances may one day enable precise predictions not only of inherited characteristics or behavioral traits, but of our responses to outside stimuli, such as medical therapeutics and drugs, leading to more targeted drug design and greatly improved healthcare.

With the emergence of this plethora of biological knowledge, some researchers feel a more quantitative approach to biology, involving mathematical modeling or computer simulations, is needed to supplement the phenomenological or empirical methods most experimental biologists traditionally use in the lab. They believe this approach might help test hypotheses to guide new experimentation, and enable theoretical insights into the behavior of complex biochemical signaling pathways in the cell. This is not to say that computational modeling of biological systems is new—since the 1960s, scientists have used mathematical models or computer techniques to study the regulation of blood pressure, metabolic processes, and immunological networks.[1] However, today's more sophisticated algorithmic techniques and greater computational power allow much more accurate modeling and more fidelity in the simulation of biological systems.

In the last several years, researchers have successfully used a circuit or "network" model to conceptualize some of the complicated forms of connectivity between genes and proteins inside cells. This approach, based on models used in electrical and computer engineering disciplines, involves viewing the molecular interactions inside the cell as a communications network of sorts composed of signaling reactions of genes and gene products with other molecules, as specific genes get "turned on," resulting in protein production. These complicated interactions underlie some of life's most fundamental processes, and are key to understanding how multicellular organisms develop from a fertilized egg.

Though living organisms all originate in a single cell—the zygote—the numerous cells that result as it divides and multiplies each assume a different function, according to which of their genes get turned on. For example, as the human zygote divides, one cell will eventually become a hand cell, and another will become a liver cell. These specialized cell functions are critically dependent on the molecular interactions that determine the coordinated expression of specific genes inside each cell. The cell organizes this activity in the form of highly complex genetic regulatory networks. Using the model of an electrical circuit or network provides an organizing principle and enables researchers to identify the causal connections between molecular reactions. In this way, they can simplify and track biochemical pathways and identify key parameters or nodes they contain.

Another advantage of using quantitative models such as these to study biological organisms is that they enable researchers to carry out computational "experiments" on the basis of the parameters the model defines. They can then compare the results from these simulations with real experimental data, which are then plugged back into the model to fine tune it. This process is usually an iterative one, with each model prediction being compared to new experimental data; modifications are then made to the model to validate it and bring it closer to approximating the actual physical system. As these models become increasingly more sophisticated and accurate, they could enable the prediction of the response of simple organisms to drugs or changes in the environment. They also represent a significant step toward being able to program a cell to act as a sensor for toxins or other harmful substances in the environment.

Though more biologists are beginning to recognize the usefulness of quantitatively based biological models, others still complain that they lack biological realism. Biology has historically been a qualitative or semi-quantitative discipline characterized by vast amounts of experimental data, so modeling and theory have traditionally played a less important role than more empirical methods.

This chapter then reverses the paradigm of biologically inspired computing that has been the basis of this book up to now. Instead of biology exerting its influence on various areas of computer science, in this case, conversely, concepts and models from information processing and engineering—such as circuits, digital logic, memory, oscillators, and switches—are applied to the investigation of biological systems, enabling novel theoretical insights and new heuristic tools for research.

The Genetic Switch of Phage Lambda

Several decades ago, biologists observed a perplexing property of a certain strain of the common intestinal bacterium *Escherichia coli*. When they exposed it to ultraviolet light, these bacteria stopped growing; yet a short time later, they burst open (or "lysed" in biological terms), spurting out a mass of viruses called lambda. These viruses are also called *bacteriophages* ("bacteria eaters"), or just *phages*. This crop of lambda phages reproduce by infecting new bacteria, many of which in turn burst open and spew out new phages. Several, however, survive by carrying lambda in a dormant form (called "lysogeny"). This group of bacteria grow and divide normally until they are exposed to UV light, at which point they, too, lyse and disgorge multiple phages[2] (figure 10.1).

Perplexed as to why one batch of infected bacteria burst and produced phages, whereas another simply kept the virus dormant, these researchers surmised that something must be acting as an on/off switch activating various genes in the bacteria, thus affecting their behavior.

In general, though each of our cells contain complete copies of our DNA or genetic material, at any given time, each cell only uses a fraction of its genes for the production of proteins. These genes are transcribed into mRNA and are said to be "expressed." Expressed genes are *turned on*, whereas those that are not expressed are *turned off*. For example, when skin is exposed to sun and gets tanned, the skin pigmentation genes themselves don't change, but they are activated or turned on by the sunlight. Biologists have since made considerable progress in understanding how this mechanism works at the molecular level, and how it is connected as an on/off switch to a larger intracellular network that regulates genetic behavior.

In a well-known paper published in 1995, a husband and wife research team at Stanford University, Harley McAdams (a systems engineer turned biologist) and Lucy Shapiro (a developmental biologist) attempted to model the switching mechanisms by which lambda phage infects its host and either causes it to disgorge more phages (lysis) or else remain dormant (lysogeny). They found that modeling this genetic regulatory network as an electrical circuit enabled them to identify the principle signaling pathways and better elucidate the workings of the switching mechanisms:

Genetic networks that include many genes and many signal pathways are rapidly becoming defined. . . . As network size increases, intuitive analysis of feedback effects is increasingly difficult and error prone. Electrical engineers routinely analyze circuits with thousands of interconnected complex components . . . [and] there are many parallels in their function [with genetic regulatory networks]. These similarities lead to the question: Which electrical engineering circuit analysis techniques are applicable to genetic circuits that comprise tens to hundreds of genes?[3]

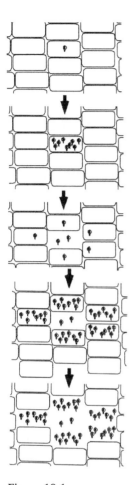

Figure 10.1
Lysis by phage lambda in a group of cells. Proceeding from top diagram down:
A phage infects a single cell in a group; phage reproduces in the cell; lysis of the
cell releases phage into the medium. Phage diffuses throughout the medium and
infects adjacent cells; phage reproduces in these cells, releasing additional
progeny; these cells lyse, releasing more phage which can then diffuse outward
and infect the surrounding cells. (Diagram courtesy of Dr. Stanley Maloy,
Department of Biology, San Diego State University)

McAdams and Shapiro found that mathematical equations describing the dynamics of chemical reactions, circuit diagrams, and computer simulation tools typically used to analyze electrical circuits could help model lambda phage behavior, especially when one wanted to predict the behavior over time of the interconnected system. To illustrate the concept, they began with a highly simplified wiring diagram of phage behavior, showing the most essential aspects of connectivity. The actual lambda phage genetic decision network is extremely complex, involving elaborate feedback loops and a byzantine complex of molecular interconnections.

Since then, additional research by McAdams, Shapiro, Adam Arkin at Lawrence Berkeley National Lab, and several others has shown that in order to be more realistic, their models must also involve stochastic (random noise) effects. Biological processes are not deterministic in nature, particularly when they involve large numbers of molecules; there are always fluctuations, adjustments, and transitions involved as these processes occur. Although an electrical circuit model has worked well to model some genetic signaling pathways, like most models it presupposes certain simplifying assumptions and cannot give a complete account of these highly complex, dynamic biomolecular systems that are constantly changing, aging, and adapting to what goes on around them.

The Molecular Workings of the Phage Switch

A look at what is actually happening at the cellular level may help explain the utility of an electrical circuit model to help elucidate lambda phage behavior. The on/off switching in lambda phage takes place due to the action of a repressor gene. When the lambda phage infects the host bacteria, its inserts its DNA into the host. The repressor gene keeps most genes turned off. If a UV light illuminates the cell, the repressor gene is ultimately digested, and lambda genes get turned on, excising their DNA from the host DNA. The phage replicates, forming new phages that are released from the host in the lysing process.

In the opposite cycle, lysogeny, all but one of the phage genes inside the host are turned off, and one phage chromosome becomes part of the host chromosome. As the host grows and divides, its DNA—including the phage chromosome—gets distributed to the daughter bacteria, and so on, for many generations—unless, of course, any one of these descendents of the original host is irradiated with UV light.

In this case, the single phage chromosome that is switched on produces a protein, the lambda repressor, which acts as both a positive and negative regulator of gene expression. It represses the action of all phage genes except one, itself, which it activates. This repressor molecule will continue turning off any additional lambda genes that are inserted into the host, effectively immunizing it against phage infections.

In the presence of UV light, the repressor molecule gets inactivated, while a second phage regulatory protein, Cro, is produced. Cro promotes lytic growth and is therefore a promoter protein. These two repressor and promotor proteins, in their interaction with the host DNA, constitute the on/off switch. With UV light, the action of these will bring about lysis—the production and spewing out of new phages—or lysogeny, the dormancy of the phage inside the host.

Synthetic Cellular Networks

In addition to the mapping of genetic regulatory networks, as in McAdams and Shapiro's work, an alternative approach consists of actually designing and constructing circuits de novo inside cells to carry out a particular function, using "off-the-shelf" biological parts from *E. coli* and its viruses. Initial efforts in this area were successfully taken by several groups, including Jim Collins and Tim Gardner et al., at Boston University, and Michael Elowitz and Stan Liebler, then at Princeton. These researchers constructed, respectively, a synthetic on/off switch and a synthetic oscillator mechanism within the DNA of living bacteria. Their work presages the growth of a cellular engineering discipline where precise logic or control systems are synthetically engineered within the cell, programming it to carry out complex instructions upon demand.

One advantage of this approach, according to Jim Collins and colleagues, is that it will aid researchers in breaking down these highly complex genetic networks into "submodules," providing a simpler template for analysis:

A central focus of postgenomic research will be to understand how cellular phenomena arise from the connectivity of genes and proteins. This connectivity generates molecular network diagrams that resemble complex electrical circuits, and a systematic understanding will require the development of a mathematical framework for describing the circuitry. From an engineering perspective, the natural path towards such a framework is the construction and analysis of underlying submodules that constitute the network. Recent experimental advances in both sequencing and genetic engineering have made this approach feasible through the design and implementation of synthetic gene networks amenable to mathematical modeling and quantitative analysis.[4]

In addition to helping to advance our understanding of complex biological systems, these synthetic genetic networks could have direct applications in biomedicine and biotechnology. For example, cells could be built to act as sensors for detecting glucose levels in the blood for diabetics, or to produce a cancer fighting protein in the presence of a cancerous tumor. In the case of those who must take a maintenance medication every day for a chronic condition—for example, hypertension, hyperthyroidism, or depression—genetic switches could potentially be engineered to be turned on by a one-time dosage of the drug. They would remain switched on to produce the drug's desired therapeutic effect until another one-time drug was taken to switch it off. This sort of transient drug delivery system could eliminate the undesired side effects resulting from taking conventional maintenance drugs over a period of years. Says Gardner, "Eventually, the goal is to produce 'genetic applets,' little programs you could download into a cell simply by sticking DNA into it, the way you download Java applets from the Internet."[5]

Toggles and Oscillators

Jim Collins, director of the Center for Biodynamics at Boston University, and Tim Gardner, then a postdoctoral researcher in his group, designed and built a molecular toggle switch—comparable to an on/off switch in a computer network—from scratch inside living cells. The two were looking to engineer a mechanism that could turn on or off a particular genetic process with the relative simplicity of a light switch. They first devised a mathematical model based on rate equations for certain intracellular chemical reactions to predict how the hypothetical switch would respond inside the cellular environment. Then they constructed a switch from genetic material in *E. coli*, using plasmids (a circular loop of DNA found inside bacterium that can be used as a vehicle for transferring DNA from one cell into another). The switch consisted of two pairs of complementary genes—promotor genes that activate genetic expression, as well as repressor genes, that inhibit expression—stacked in such a way that only one of the promotors could be active at a time. An external stimulus was needed to allow either one or the other gene to take over— they used either a synthetic analog of sugar or tetracycline, or a rise in temperature. They also built a green fluorescent protein into their switch that would glow under UV light when the switch was on, to enable them to monitor the system. They found that when they used the right configuration of genetic material under the right conditions, the flip from one switch to another was stable, even after the effect of the initial stimulus or temperature rise was gone. Collins and Gardner called their artificial switch a "genetic applet" (figure 10.2).

They also discovered that their mathematical model was predictive only up to a point, as it had been constructed to model the behavior of the "average" cell, and did not account for fluctuation, noise, or stochastic effects in genetic behavior. As a result, they observed variations in the switching time as they monitored the system. More stable system behavior would be needed to make the switch useful for biomedical applications.

Michael Elowitz and Stan Leibler, then at Princeton University, wanted to build a synthetic system inside the cell that would act to inhibit or repress the function of certain genes. They put four genes into a plasmid

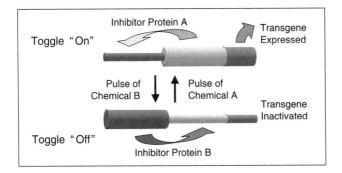

Figure 10.2
Design and operation of the genetic toggle switch, which consists of two pairs of mutually inhibitory, complementary genes: promoter genes and repressor genes. Both transcription units may be coupled to a green fluorescent transgene, which indicates when the switch is "on" or "off." (Image courtesy of Tim Gardner)

and inserted it into *E. coli*. Each of the first three genes repressed the activity of the next. The fourth gene controlled the production of a fluorescent protein, so that pulses of the fluorescence indicated the periodic output of the oscillator or clock. They called their system "the repressilator."

The system cycles took about an hour and a half for each cycle, and the oscillations lasted for ten cell cycles. The scientists had designed the network according to a mathematical model, and although some cells conformed to the model behavior, others did not. Again, the effects of noise, fluctuations, and stochastic processes showed the variable nature of biological systems.

More scientists are now carrying out similar efforts to model, analyze, and even engineer artificial genetic networks inside living cells. Their work serves as proof of the principle that at some level cellular communication can potentially be not only understood and mapped, but programmed from the outside to carry out simple actions, such as switching, oscillation, or even a simple digital logic gate. However, we are still a long way from being able to accurately predict the behavior of a genetic regulatory circuit, albeit very simple ones.

Microbial Engineering

It takes only a little imagination to go from the idea of synthetic switches or oscillators in a genetic regulatory network to being able to program cells in the lab to carry out prespecified behavior. Computer scientist Tom Knight of MIT's AI Lab, who played a major role in the field of computer architecture and engineering (see chapter 7) having designed one of the first computer work stations, says he was led to biocomputing by realizing that silicon-based computing would eventually come to an end, and that we needed viable alternatives.

"At that point, I started thinking hard about doing molecular-scale assembly of integrated circuits," recalls Knight.

However, the piece that pushed me over the edge was an article by biologist Harold Morowitz, in the late '80s, which basically said that it was possible to reach a complete understanding of single-celled organisms. The article had a profound affect on me, and made me realize that one day, we could actually engineer living systems. At this critical point in my thinking, I felt, OK, let's just go do this. So I started doing a lot of reading in biology, and took some MIT graduate bio courses. Several years ago, I started learning bench biology, and have been in the lab doing experiments ever since.[6]

Knight admits to having "flailed around a lot" at first in the wet lab, "but the upshot is that biology isn't black magic for me anymore. I've made all the mistakes, been through it all, and there's no substitute for that."[7] Observing daily the workings of biological organisms at first hand has enabled him to "learn things and transport them back to the computer science realm."[8] Although the Collins/Gardner switch and Elowitz/ Liebler oscillator have provided an initial proof of concept for Knight's vision, his overarching goal is to be able to program bacteria the way a computer scientist might program a microprocessor.

Together with former grad student Ron Weiss, they constructed the equivalent of a logic gate in a digital circuit in a cell's genetic regulatory network. The logic signal itself is represented by the rate of production of certain proteins in the circuit. These proteins can function as repressors—rather than promotors—of genetic activity, so the effect of a repressor protein on the production of another would constitute either a one (no inhibition of protein production) or a zero (repression of protein creation), and so forms the basis of the digital inverter. (An inverter is a type of logic gate that inverts the signal it receives, that is, it converts a one into a zero and vice versa. It is also called a NOT gate. See figure 10.3.) The group has also been able to contruct more complicated gates by combining inverters.

Weiss, now a faculty member in Princeton's Electrical Engineering Department, has made significant advancements in this direction, building other types of intercellular logic gates and a five-gene circuit in *E. coli* that can detect a particular nearby chemical and emit a signal when its concentration reaches a certain level. This could have potential environmental or even biomedical applications for detecting toxins or other unwanted chemicals. He is also experimenting with intracellular circuits that "evolve," letting natural selection carry out additional steps. By letting the DNA involved in the circuit mutate, he has discovered that, in one case, the mutated DNA circuits performed better than the original circuit.

Knight and present collaborator Drew Endy of MIT's Biology Department have recently formed a research group for "synthetic biology," with the goal of creating "a general engineering and scientific infrastructure

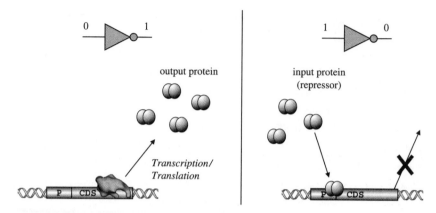

Figure 10.3
The biochemical reactions that occur when a gene is expressed can be used to implement a digital logic signal. In the state depicted on the left, when the input repressor protein is absent, RNA polymerase, needed for gene expression, is free to bind DNA at the promotor site and express the output protein, resulting in a one, digitally speaking. On the right, when the input repressor protein is present, RNA polymerase can no longer bind DNA at the promotor site, and the digital output is zero. (Image courtesy of Ron Weiss, Princeton University)

that enables the design and construction of large-scale synthetic biological systems."[9] Their point of departure is that cells are already wonderfully efficient chemical factories, naturally producing antibiotics, enzymes, and other useful chemicals. They can detect minute changes in their environment and respond accordingly. With the goal of harnessing these natural abilities of cells, they also want to clearly define system requirements in terms of biological components and performance, characterize biological parts and devices, develop methods and tools that support system design and programming, and create improved biofabrication methods.

Their current projects include rebuilding an organism's genome and establishing a library of standardized off-the-shelf biological parts or components, called "BioBricks"—all highly challenging tasks that "push the envelope" on what can currently be clone in this field. Cellular circuitry itself is so complex and still so poorly understood that the BioBricks must use those few genetic mechanisms that are simple enough to produce the desired proteins, and avoid those that could interfere with the circuit.

The Spice of Life

The integrated circuit (IC) used in microelectronics is probably one of the most important inventions of the twentieth century—serving as the basis for all of today's computer, electronics, and communications devices. Made up of infinitesimally small electrical circuits placed in an intricate pattern on a silicon chip, smaller than a grain of rice, early microchip designers needed a way to predict the electrical characteristics of the chip before manufacturing it. In the 1970s, researchers at the University of California at Berkeley developed a computer program named "Simulation Program with Integrated Circuit Emphasis" (SPICE) to simulate the behavior of complex electrical circuits, particularly ICs. It enabled the user to model the spatiotemporal behavior of any given integrated circuit containing elements such as resistors, capacitors, or inductors, including the most common types of semiconductor devices. SPICE became so widely adopted that it eventually became an industry standard; it is still used the world over today, in addition to serving as a highly effective university teaching tool.

The biological analogue of SPICE, BioSPICE (a Simulation Program for Intra-Cell Evaluation), was recently developed by researchers with funding from DARPA. Sri Kumar, the project's program manager, felt that an open source, user-friendly simulation tool was needed to enable researchers to try and map a variety of intracellular processes, including genetic signaling networks or cell division—for exploratory as well as practical ends, such as gene therapy. By using expertise from biologists as well as computer scientists, these simulations aim to explore how environmental changes, such as temperature or acidity, affect these networks, as well as how various pathogens act to disrupt normal cellular processes. BioSPICE models could also drive innovations in the design of de novo synthetic networks and network components.

BioSPICE version 1.0 was developed to handle some of the simpler, better understood network models; subsequent versions will be able to model higher order aspects of cellular networks, such as feedback loops, non-linear interactions, and stochastic effects. The final version of BioSPICE is intended to operate on various levels—molecular, biochemical, or genetic—with varying degrees of resolution (figure 10.4). One key aspect of BioSPICE is that it is being constructed in a

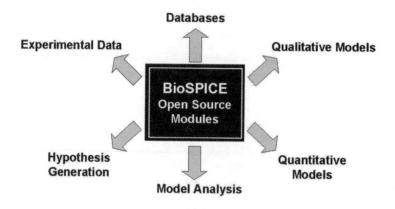

Figure 10.4
DARPA's BioSPICE Program has developed a comprehensive suite of software tools for creating and exploiting sophisticated models of intracellular processes. These open source software modules can: (1) provide access to databases containing raw biological data; (2) allow collaborative groups to organize and share the metadata surrounding a model or experiment; (3) help catalogue related literature; (4) generate qualitative descriptions of interacting biochemical entities; and (5) simulate biological processes quantitatively, and validate their predictions in the laboratory. (Source: Sri Kumar, Defense Advanced Research Agency)

collaborative environment, and ultimately will function as a collaborative tool among multiple researchers, who can choose to store any of their BioSPICE results in a database for use or study by others. An open source software project, its licensing agreement has been set up to promote eventual commercialization of proprietary versions of BioSPICE.

Implicit in the development of BioSPICE was the idea that wet lab experiments could validate and refine the models. Though the user may start out using any values he or she wants for some of the model's required parameters, with each run, the model will produce a set of realistic numerical values for these parameters that can then be tested against experimental values. The latter can be plugged back into the model in an iterative process, to produce resulting numbers of increasingly greater degrees of accuracy.

In actuality, the process begins with the user choosing a BioSPICE program, such as "Pathway Builder," a visual network modeling program. (BioSPICE also contains large databases and servers that store and provide access to biological data contained in national databases, such as the National Center for Biotechnology Information at the National Institutes of Health in Bethesda, Maryland.) The user then chooses the appropriate starter or precursor molecules in the database, the enzymes needed to catalyze the biochemical reactions, and the intermediate molecules and reactions—all of which will make up the network. Once she or he has built the pathway, the user inputs it into Pathway-Builder, enters the required values for parameters—such as system temperature, initial concentration of various chemicals, and so on—and finally runs the simulation. The final model can be stored, edited, or finetuned by continued comparison with experiment.

Digitally Inspired Biology

Could any of these developments have occurred without the aid of computers serving as metaphor, enabler, or simply inspiration in a multidisciplinary research setting? Possibly, but chances are they would have occurred in a much more hit-or-miss fashion and at a considerably slower rate. It's doubtful that by using purely empirical experimental methods, researchers could acquire the systems perspective, borrowed from engineering, that enables them to see the whole as more than the sum of its parts, and work with it accordingly. In addition, computer simulations have enabled these researchers to test out hypothetical cellular mechanisms, and then return to the wet lab to validate or refine them.

By approaching intracellular signaling conceptually as a communications network, scientists have been able to draw from the storehouse of knowledge and tools built up by computer scientists and systems engineers to guide them in their investigations and facilitate novel developments such as Gardner's toggle switch, Elowitz's oscillator, Weiss's logic gates, and Knight and Endy's "synthetic biology."

These developments hold tremendous promise for applied biomedicine or environmental uses, among other areas. They could eventually help us to deliver drugs more efficiently, or travel through the bloodstream and monitor blood sugar for diabetics, manufacture natural antibiotics, or even detect the faint beginnings of a cancerous growth in time to remove it. It must be said, however, that there is also a potential dark side to man-made advances such as these and others in biotechnology. The possibility looms not only of their unethical or unauthorized use, but also the unthinkable application of these advances as some form of biowarfare in the hands of rogue nations or terrorist groups. It remains to scientists involved in this area to develop an awareness of "the dark side" of biology in today's world, and anticipate and communicate any potential for misuse, while working with others to ensure as best they can that these important advancements are not used for ill purposes.

Epilogue

Natura Generatrix
—Latin Proverb

Information processing systems today have become ubiquitous, so embedded in the majority of machines and processes that populate our daily lives that at times we can become almost oblivious to their presence; yet our dependence on them grows every day. Moreover, as our computing systems, with the elaborate web of networks that connect them, evolve to become more sophisticated technologically, with greater speed and span of control, they grow correspondingly more complex in design, operation, and capabilities, such that to most people, they remain nothing more than a web of mysterious black boxes.

Not only the machines themselves, but the tasks they are required to perform are also growing increasingly more complicated, in terms of their scope, degree of organization and detail, the extensive computer codes and data inputs they require, and the need for extreme precision in their implementation. One need only consider the extent of the entire computing system that oversees and manages the space shuttle's activities during launch and in orbit, for example, or the intricate computer software needed to run the abstract mathematical models that underlie our theories of global climate change, the evolution of galaxies, or the propagation of a nuclear explosion through solid matter to illustrate this trend for increasing degrees of complexity and detail in our information processing systems and programs.

As a result, many computer science researchers and IT professionals have expressed concern about the consequences if one of these elaborate systems fails. The more complicated the system, the greater its range of control, and the more critical its functions, the more devastating an error, malfunction, or breakdown could be. Faults can occur not only at the systems level, but defective hardware, bugs in the software, changes in the operating environment, and the threat of cyberattacks all cause today's computing systems to be highly vulnerable, and as a consequence, put large portions of society at risk.

Though varying levels of fault tolerance and redundancy can be and *is* incorporated into modern computing systems for reasons of safety (the space shuttle has as many as five backup computers), as John von Neumann so presciently realized in the 1940s, our computer designs, as a rule, tend to be so inflexible and unwieldy that one small error in a billion-step computing algorithm can completely invalidate the end result. He believed this put far too much emphasis on the need for precision and reliability in computers—a requirement that runs inversely proportional to the trend for growing complexity.

Von Neumann also had the acumen to recognize that biological systems, once we came to more fully understand their fundamental organizational principles and behavior, could be a wellspring of inspiration, capable of causing a gestalt switch in our thinking about machine malfunctions, enabling them to be more resilient, to better adapt to change, to self-diagnose, and even to self-heal. He observed how human beings can continue to function despite the presence of a nonfatal illness or injury, because nature has engineered the body to have inherent redundancy, resiliency, and plasticity and an extraordinarily efficient immune system that can identify, fight, and destroy the anomaly to bring about healing. These biological traits have been optimized through natural selection as organisms incrementally improved their species' ability to survive under various stresses such as changes in temperature, pressure, climate, food intake, predation, and habitat.

In sum, learning to endow computing systems with greater degrees of adaptability, fault-tolerance, redundancy, and ability to self-heal is only one of the many lessons we can extract from the study of natural automata, to echo von Neumann, and apply to the design of artificial automata. As I hope this book has illustrated, there are indeed many other valuable concepts and models inherent in the bio-inspired paradigm that can be applied to computing, such as the ability to structure an algorithm so that it can learn the way our brains learn, and be able to recognize patterns; optimization through the principles of Darwinian natural selection; local actions giving rise to global phenomena; cellular differentiation and morphogenesis as a template for information processing via amorphous computing; algorithms modeled on tropisms or via physical strands of DNA; co-opting photosynthetic mechanisms for information storage or energy production; and others. More and more, scientists and engineers are becoming aware that, in their search for new approaches to some of today's information processing conundrums, or simply novel avenues for research, they can look to natural systems for inspiration, ideas, originality, and even solutions. They can identify and extract the relevant biological notions and apply them to computing, finding appropriate ways to express or realize such biomimetic concepts as growing point, symbiosis, ecosystem, learning, or ontogeny in the language of information processing.

The bio-inspired fields included in this book have been and will continue to be stimulated by the knowledge explosion currently taking place in the life sciences, as researchers persist in pushing forward the frontiers of knowledge in their understanding of the phenomenology of life—in particular, in our knowledge of protein function and of the human brain. Developments in instrumentation, such as gene sequencing machines, high throughput measuring devices, and mass spectroscopy, in biological imaging devices, in experimental techniques, such as PCR, microfluidics, and microarrays, and in our ability to manage and store biological data (known as bioinformatics) have all been key drivers for the rapid advancement of life sciences research. Our knowledge that genetic material was made up of DNA, and that biological organisms had a blueprint consisting of finite numbers of genes has led us to the fundamental insight that the sequence of these genes encoded all the information required to specify the reproduction, development, and adult function of an individual organism. These advances have also given rise to many new subdisciplines in the life sciences, such as genomics (the study of gene sequences in living organisms and the ability to read and interpret them), proteomics (the identification of proteins and their function in the body), structural biology (uncovering the structure of proteins), and others, flush with new students, practitioners, and research dollars, while spinning off many useful applications in biotechnology, biomedicine, and biodefense.

The knowledge boom in the life sciences has also served to inspire and inform other disciplines, in addition to computer science, with an affinity for biomimetics, such as engineering and economics. More than just an external influence, however, biology can also act in creative synergy with these fields. In a continual positive feedback loop, computer scientists and other nonbiologists working in interdisciplinary bio-inspired fields have been able to provide, in their turn, a unique perspective on the study of biological systems. For example, systems engineers and those trained in control theory are finding they can successfully apply certain methods to genetic or cellular behavior, resulting in notions such as biological networks, cellular reengineering, or the novel concept of systems biology, which examines the total structure and dynamics of cells and organs, rather than the properties of their isolated parts. This development is enabling scientists to see characteristics of the system as a whole, such as robustness or modularity, which would not appear otherwise.

A group of scientists and technologists at a recent World Economic Forum in Davos, Switzerland declared that technology created from the fusing of computer science and biology may likely be the most important trend to emerge in the future, possibly even dwarfing the importance of the Internet. Regardless of whether their prognostications prove true, the marriage of these two disciplines has already proved a fruitful one, having produced significant developments in areas such as algorithms, hardware, devices, distributed sensing, and information security, among others. Their merger has also given rise to several promising new multidisciplinary fields of research and attracted some of the best and brightest in science, drawn by the fascination of Nature's engineering, and the creative symbiosis formed by the melding of disciplines, the cutting-edge nature of the work, the potential for applications, and the challenge of searching for biomimetic solutions to difficult problems in computer science. As life sciences research continues to burgeon with new discoveries and understanding, possibly (as experts claim) leaving as strong an imprint on the twenty-first century as physics left on the previous one, there's little doubt that it will also continue its sustained interplay with the information sciences, with the potential for enormous benefits in both science and its practical applications.

Notes

Preface

1. L. Jeffress, ed., *Cerebral Mechanisms in Behavior: The Hixon Symposium.* New York: John Wiley & Sons, Inc., 1951, pp. 2–3.

2. According to the *Oxford Dictionary of Computing*, 4th ed., a von Neumann machine is characterized by having a central processor, memory, input, and output functions; programs and data that share the same memory, making the concept of a stored program fundamental; and a central processor that determines the actions to be carried out by reading instructions from the memory.

3. T. Sterling, P. Messina, and P. Smith, *Enabling Technologies for Petaflops Computing.* Cambridge, MA: MIT Press, 1995, p. 15.

Chapter 3

1. W. Aspray, *John Von Neumann and the Origins of Modern Computing.* Cambridge, MA: MIT Press, 1990, pp. 189–190.

2. L. Jeffress, op. cit., p. 1–2.

3. Ibid. pg. 23.

4. J. Casti, *Reality Rules: I.* New York: John Wiley & Sons, Inc., 1992, p. 216.

5. S. Levy, *Artifical Life.* New York: Vintage Books, 1992, p. 29.

6. S. Weinberg, "Is the Universe a computer?" *The New York Review of Books,* Vol. 49, No. 16, 2002, pp. 8–10.

Chapter 4

1. Personal communication with John Casti, January 2001.

2. C. G. Langton, *Artificial Life: An Overview.* Cambridge, MA: MIT Press, 1995, p. ix.

3. Ibid., pg. 113.

4. K. Sims, "Evolving 3D Morphology and Behavior by Competition." In *Proc. Artificial Life IV*, R. Brooks and P. Maes, eds., Cambridge, MA: MIT Press, 1994, pp. 28–39.

5. C. Adami, *Introduction to Artificial Life*. New York: Springer-Verlag, 1998, p. 13.

6. J. L. Casti, *Would-be Worlds: How Simulation in Science Is Changing the Frontiers of Science*. New York: John Wiley & Sons, 1997, p. 165.

7. D. E. Nilsson and S. Pelger, "A Pessimistic Estimate of the Time Required for an Eye to Evolve," *Proc. Royal Society London B*, London 1994, pp. 53–58.

8. Personal communication with Rob Knight, June 2001.

9. Personal communication with Andy Ellington, June 2001.

Chapter 5

1. Personal communication with Len Adleman, May 2002.

Chapter 6

1. G. Whitesides, J. Mathias, and C. Seto, "Molecular Self-Assembly and Nanochemistry: A Chemical Strategy for the Synthesis of Nanostructures," *Science* 254, 5036, 1991, pp. 1312–1319.

Chapter 7

1. L. Jeffress, op. cit., p. 19.

2. Written communications with Gerry Sussman over various periods in 2000.

3. Personal communication with Hal Abelson, June 2000.

4. Written communication with Gerry Sussman, June 2000.

5. J. R. Heath, P. J. Kuekes, G. S. Snider, and R. S. Williams, "A Defect-Tolerant Computer Architecture: Opportunities for Nanotechnology," *Science* 285, 5426, 1999, pp. 391–394.

6. Personal communication with Hal Abelson, June 2000.

7. Personal communication with Tom Knight, June 2000.

8. Ibid.

Chapter 8

1. Personal communications with Jeff Kephart, January–February 2003.

2. Ibid.

Chapter 9

1. R. B. Birge et al., "Biomolecular Electronics: Protein-Based Associative Processors and Volumetric Memories," *J. Physical Chemistry B*, 8, 103, pp. 10746–10766.

2. Ibid.

3. Personal communication with Neal Woodbury, November 2000.

4. Personal communication with Devens Gust, January 2001.

5. Personal communication with Michael Kozicki, December 2000.

6. M. Sipper, D. Mange, and E. Sanchez, "Quo Vadis Evolvable Hardware," *Communications of the ACM*, 42, 4, April 1999, pp. 50–56.

7. Personal communication with Moshe Sipper, January 2001.

8. X. Yao, "Following the Path of Evolvable Hardware," *Communications of the ACM*, 42, 4, April 1999, pp. 47–49.

9. Ibid. p. 49.

10. Personal communication with Jose Muñoz, February 2001.

11. T. Higuchi, and N. Kajihara, "Evolvable Hardware Chips for Industrial Applications," *Communications of the ACM*, 42, 4, April 1999, p. 61.

12. Ibid. p. 62.

13. Ibid. p. 63.

14. Personal communication with Andre De Hon, January 2001.

15. Ibid.

16. Personal communications with Adrian Stoica, May 2002.

17. Personal communication with Andy Tyrell, March 2001.

18. Personal communications with Daniel Mange, December 2000–April 2001.

19. See note 17.

20. See note 16.

21. See note 17.

22. Personal communication with Kwabena Boahen, August 2003.

Chapter 10

1. National Academy of Sciences, *Models for Biomedical Research*. Washington, DC: National Academy Press, 1985, pp. 155–167.

2. M. Ptashne, *A Genetic Switch*. Cambridge, MA: Cell Press and Blackwell Science, 1992, pp. 13–16.

3. H. H. McAdams, and L. Shapiro, "Circuit Simulation of Genetic Networks," *Science* 269, 650, 1995.

4. J. J. Collins, J. Hasty, and D. McMillen, "Engineered Gene Circuits," *Nature* 420, 14 November 2002, p. 224.

5. Personal communication with Tim Gardner, January 2003.

6, 7, 8. Personal communication with Tom Knight, June 2002.

9. Personal communication with Drew Endy, February 2003.

Index

Abelson, Hal, 84, 86–87, 90–91, 93
Adami, Chris, 42, 44, 47–48
Adaptability, 40
Adaptation in Natural and Artificial Systems (Holland), 16
Adenine, 54–58
Adleman, Leonard
 DNA computation and, 53–54, 58–62, 65
 self-assembly and, 79
Adsorption, 75
AIDS, 53, 102
Algorithms
 amorphous computing and, 83–95
 a priori programming and, 9
 artificial neural networks and, 1–11
 automata and, 28–36
 backpropagation, 4, 8
 bio-operators and, 16–17
 Biowatch and, 131–132
 Church-Turing hypothesis and, 28
 collision avoidance, 18–20
 computer immune, 106–108
 digital Darwinism and, 14–15
 DNA computation and, 51–65
 evolutionary, 13–24
 evolvable hardware and, 122–129
 FPGAs and, 89–90, 125–132
 Game of Life and, 32–33
 gameplay and, 20–22
 genetic, 15–24, 41
 Growing Point Language and, 91–95

morphogenesis and, 91
parse tree and, 17–18
pattern recognition, 8–9
perceptrons and, 3
self-assembly and, 77–79
speech, 9
Teramac and, 90–91
Turing machine and, 27–28
Amorphous computing
 Abelson, Hal, and, 84, 86–87, 90–91, 93
 fault tolerance and, 83–85, 88–91
 Growing Point Language and, 91–95
 hardware experiments in, 85–88
 Knight and, 91, 93, 95
 logic elements and, 87
 morphogenesis and, 91
 smart paint and, 85–86
 software and, 91–95
 Sussman and, 84, 87, 93
 Teramac computer and, 88–91
Annealing, 56
Antigens, 99
Arizona State University, 120–122
Art, evolutionary, 22–24
Artificial intelligence (AI), 42, 148–150
 cellular automata and, 29
 neural networks and, 10–11
Artificial life
 adaptability and, 40
 Avida program and, 47–48

Artificial life (cont.)
bottom-up approach to, 41–42
emergent behavior and, 44–46
GOLEM project and, 42, 44
individual entities and, 41–44
natural selection and, 40
non-carbon-based, 38–40
origins of, 40–41
real life and, 48–50
robots and, 42
tenets of, 37–38, 48–50
Tierra project and, 46–47
Associative memory devices, 116–118
Automata, xi–xii, 26, 29–31
applications of, 34–36
biological system logic and, 28–29
cellular, 31–36
definition of, 26
fault tolerance and, 83
genetics and, 31
kinematic self-replicating, 29–31
Turing and, 27–28
von Neumann and, xi–xii, 25–26,
28–31
Avida program, 47–48
Axons, 4–5

Backpropagation algorithm, 4, 8
Bacteriophages, 141–146
Bacteriorhodopsin, 114–115, 120
associative holographic memory
and, 115–118
optical memory and, 118–120
B cells, 102–103
Behavior, 2
artificial life and, 37–50
automata and, 25–36
emergent, 44–46
evolvable hardware and circuit,
128–129
Tierra project and, 46–47
BioBricks, 150
Biohardware
associative holographic memory
and, 116–118
biomaterials and, 113–114

Biowatch and, 131–132
data storage and, 114–115
electronics and, 113–114
embryonic, 129–131
energy and, 121
evolvable, 122–129
immunotronics and, 133–135
marine animals and, 138
neurally inspired, 135–138
optical memory and, 118–120
photosynthesis studies and, 120–122
transistors and, 121, 137
volumetric memory devices and,
115–116
Biology, ix
artificial life and, 37–50
artificial neural networks and, 1–11
automata and, xi–xii, 25–36
biocomputation and, 77–79
biohardware and, 113–138 (*see also*
Biohardware)
Church-Turing hypothesis and, 28
ciliates and, 64
computers and, ix–xiii, 153–154
Darwinism and, 13–15
DNA computation and, 51–65
embryogenesis and, 129–131
evolutionary algorithms and,
13–24
fault tolerance and, 83–85, 155–156
Game of Life and, 32–33
genetic switches and, 141–149 (*see
also* Genetics)
Growing Point Language and,
91–95
Hixon Symposium and, x–xii, 26
immune systems and, 98–104
increasing knowledge of, 139–141,
155–158
intracellular signaling and, 139–148,
153–154
as metaphor, xiii
microbial engineering and, 148–150
modeling and, 140–141
oscillators and, 146–148
pigmentation and, x

random noise and, 144
self-assembly and, 67–81
self and nonself and, 98–103,
105–106
synthetic, 149–150, 153
system logic for, 28–29
Turing and, x
von Neumann and, x–xii
zygote and, 140–141
BioSPICE, xiv, 151–153
Biowatch, 131–132
Birge, Bob, 114–120
Boahen, Kwabena, 136–137
Boids, 45
Brain, x–xiii, 156–157
artificial neural networks and, 1–11
axons and, 4–5
biohardware and, 135–138
dendrites and, 4–5
Fourier transform association and,
116
mathematics and, 1–2
processing mechanisms of, 4–5
synapses and, 5
von Neumann's computers and, 1
Brown, Titus, 47
Burks, Arthur, 40–41

Calculating machines, 1
Casti, John, 35, 39
Cellular Automata Machine (CAM),
32–34
Chellapilla, Kumar, 20–21
Chemistry, x, 74–75
artificial life and, 38–40
biocomputation and, 77–79
self-assembly and, 72–73
Chess, 20
Church-Turing hypothesis, 28
Ciliates, 64
CMOS technology, 93, 137
Cold War, 115
Collins, Jim, 145–147, 149
Computational particles, 85–86
Growing Point Language and,
91–95

Computer immune systems
autonomy in, 104
biohardware and, 133–135
biology and, 98–103
distributability in, 103, 107–108
diversity in, 104
dynamic coverage and, 104
encryption and, 104–105
fault tolerance and, 103
Forrest, Stephanie, and, 101,
104–108
intruder detection and, 105–107
memory and, 104
multiple layers and, 104
novelty detection and, 104
self and nonself and, 105–107
virus hunting in, 108–111
Computer science
amorphous computing and, 83–95
a priori programming and, 9
artificial life and, 37–50
artificial neural networks and, 1–11
Avida program and, 47–48
biocomputation and, 77–79
biohardware and, 113–138 (*see also*
Biohardware)
cellular automata and, 32–36
central processing unit (CPU) and, 9
Church-Turing hypothesis and, 28
Cray, Seymour, and, xii–xiii
Deep Blue and, 20
digital Darwinism and, 14–15
DNA computation and, 51–65
encryption and, 53, 62–63, 104–105
evolvable hardware and, 122–129
fault tolerance and, 83–85, 88–91
FPGAs and, 89–90, 125–126, 128,
132
Game of Life and, 32–33
gameplay and, 20–22
genetic algorithms and, 15–24
Growing Point Language and,
91–95
immune systems and, 97–111
increasing knowledge of, 139–141,
155–158

Computer science (cont.)
  neural networks and, 9–10 (*see also*
    Neural networks)
  optimization and, 17
  satisfiability problem and, 61–62
  SIMD computation and, 56
  system failure concerns and,
    155–156
  Teramac computer and, 88–91
  theoretical foundations of, ix–x
  Tierra project and, 46–47
  Turing machine and, 27–28, 80
  viruses and, 97, 106–111, 133–135
Consortium fur elektrochemische,
    Industrie GmbH, 120
Conway, John Horton, 32–33
Coore, Daniel, 91–93
Cray, Seymour, xii–xiii
Crick, Francis, xi, 31
Cro protein, 145
Crossover operator, 16
Cyber attacks
  biohardware and, 133–135
  computer immune systems and, 97,
    106–111, 133–135
Cytosine, 54–58

DARPA, 151
Darwin, Charles, 13–15
Data encryption standard (DES), 62
Data storage. *See* Memory
Dawkins, Richard, 22
Deep Blue, 20
DeHon, Andre, 128–129
Denaturation, 56
Dendrites, 4–5
Differential equations, 35
Digital switching, xii
Diorio, Chris, 136–137
Diploid organisms, 17
Directed Hamiltonian Path Problem,
    53–54, 58–61, 79
Ditto, William, 138
DNA, ix, xi, 157
  automata and, 31, 35
  double helix of, 54

evolutionary algorithms and, 14–15
  genetic switches and, 141–149,
    153
  molecular structure of, 51, 54–58
  problem solving and, 51–52
  self-assembly and, 52, 71, 75–76,
    79–80
DNA computation
  Adleman and, 53–54, 58–61, 65
  ciliates and, 64
  Directed Hamiltonian Path Problem
    and, 53–54, 58–61, 79
  gene-based computer and, 54–58
  Head, Tom, and, 53
  limitations of, 52
  Lipton, Dick, and, 61–63
  NP-completeness and, 60–61
  parallelism of, 62–63
  satisfiability problem and, 61–62
  skepticism toward, 63–64
  splicing model and, 53
Dynamic systems
  amorphous computing and, 83–95
  artificial life and, 37–50
  cellular automata and, 32–36
  computer immune systems and,
    97–111
  emergent behavior and, 44–46
  evolvable hardware and, 122–129
  MEMs and, 85, 114
  self-assembly and, 67–81
Dyson, Freeman, xiv, 31

Electron microscopy, 67, 86
Electrostatics, 72
Ellington, Andy, 49
Elowitz, Michael, 145, 147–149,
    153
Embryonic hardware, 129–131
Emergent behavior, 44–46
Encryption, 53, 62–63, 104–105
Endy, Drew, 149–150
Epitopes, 99
Error derivative, 8
Error derivative of the weight, 8
Escherichia coli, 141–142, 145–149

Evolution
  art and, 22–24
  artificial life and, 38–39 (*see also*
    Artificial life)
  digital Darwinism and, 14–15
  genetic algorithms and, 15–18
  as metaphor, 13–14
  natural selection and, 13–15, 18–20,
    40, 103, 157
  Tierra project and, 46–47
Evolvable hardware (EHW), 122
  applications of, 127–128
  immunotronics and, 133–135
  intermittent behavior of, 128–129
  mechanisms of, 125–127
  real world problems and, 123–124
  reasons for using, 124–125
Excitatory signals, 5

Fault tolerance, 155–156
  amorphous computing and, 83–85,
    88–91
  computer immune systems and, 103
  embryonic hardware and, 130–131
  immunotronics and, 133–135
  Teramac computer and, 88–91
Field programmable gate arrays
    (FPGAs), 89–90, 125–126, 128,
    132
First Response software, 108
Fogel, David, 20–21
Fogel, Lawrence, 15
Forrest, Stephanie, 101, 104–108, 135
Fourier transform, 116

Game of Life, 32–33
Games, 20–22
Gardner, Tim, 145–147, 149, 153
"General and Logical Theory of
    Automata, The" (von Neumann),
    xi–xii, 26
Genetic applet, 147
Genetic programming, 15–18
Genetics, 35, 157
  algorithms for, 15–24, 41, 141–149,
    153

automata and, 31
BioSPICE and, 151–153
Biowatch and, 131–132
digital Darwinism and, 14–15
DNA computation and, 51–65
human genome and, 139
mRNA and, 71, 142
oscillators and, 146–148
programming and, 15–18
switches and, 141–149, 153
Golem (Genetically Organized
    Lifelike Electro Mechanics) project,
    42, 44
Growing Point Language (GPL),
    91–95
Guanine, 54–58
Gust, Devens, 121–122

Hackers
  biohardware and, 133–135
  computer immune systems and, 97,
    106–111, 133–135
Haploid organisms, 17
Head, Tom, 53
Heath, James, 89
Hebbs, Donald, 2
Higuchi, Tetsuya, 128
HIV, 53, 102
Hixon Symposium, x–xii, 26
Hofmeyr, Steve, 108
Holland, John, 16–17, 41
Holographic memory, 115–118
Hopfield, John, 3–4
Hybridization, 56

IBM, 20, 108, 111
Immunotronics. *See* Computer
    immune systems
Information, ix–x
  artificial neural networks and,
    1–11
  automata and, xi–xii, 31
  biohardware and, 135–138
  BioSPICE and, 151–153
  computer immune systems and,
    97–111

Information (cont.)
  DNA computation and, 51–65
  intracellular signaling and, 139–148,
    153–154
  recurrent networks and, 3–4
  ubiquitous systems and, 155
Inhibitory signals, 5
Input/output, 1–2
  memory and, 114–120
  neuronal, 4–8
  weighted signals and, 5–8
Integrated circuits (ICs), 85, 151–153
"Interdisciplinary Workshop on the
  Synthesis and Simulation of
  Artificial Life," 41
Inversion operator, 16
Inverter, 149

Jefferson, David, 40

Kajihara, Nobuki, 128
Kasparov, Gary, 20
Keating, Christine, 75–76
Kephart, Jeffrey, 104, 108–111
Kinematic self-replicating automaton,
  29–31
Knight, Rob, 48, 50
Knight, Tom, 84, 91, 93, 95,
  148–150
Koza, John, 17–18
Kozicki, Michael, 121–122
Kuekes, Phil, 89

Lambda viruses, 141–146
Landweber, Laura, 62, 64
Langton, Chris, 39–41
Language theory, 53
Lasers, 116–120
Lashley, Karl, x
Layzell, Paul, 129
Lenski, Richard, 47–48
Liebler, Stan, 145, 147–149
Lipton, Dick, 61–63
Logic, ix
  amorphous computing and, 83–95
  art and, 22–24

artificial life and, 40
artificial neural networks and,
  1–11
automata and, xi–xii, 25–36
biological systems and, 28–29
Church-Turing hypothesis and, 28
DNA computation and, 51–65
evolvable hardware and, 122–129
fault tolerance and, 83–85
FPGAs and, 89–90, 125–126, 128,
  132
gates, 149, 153
genetic switches and, 141–149, 153
Growing Point Language and,
  91–95
perceptrons and, 3
Traveling Salesman Problem, 53–54,
  58–61, 79
Turing machine and, 27–28
"Logical Calculus of the Ideas
  Immanent in Nervous Activity, A"
  (Pitts & McCullough), 1–2
Lymphocytes, 102–103, 107
Lysogeny, 142, 144–145

McAdams, Harley, 142, 144
McCullough, Warren, x, xii, 1, 3, 10,
  27–28
Macrophages, 99–100
Mallouk, Tom, 75–76
Mange, Daniel, 130
Margolis, Norman, 33–34
Mathematics, x
  Adleman and, 53–54, 58–61
  artificial life and, 37
  brain processing and, 1–2
  differential equations, 35
  Fourier transform, 116
  genetic switches and, 144
  modulo arithmetic, 132
  Navier-Stokes equations, 35
  neural networks and, 1–3
  NP-completeness and, 60–61, 129
  perceptrons and, 3
  threshold value and, 2
  tilings, 80

Traveling Salesman Problem, 53–54, 58–61, 79
Turing machine and, 27–28
Mead, Carver, 136–137
Meissner effect, 67–68
Memory, 1–2
  associative, 116–118
  bacteriorhodopsin and, 114–115
  Fourier transform and, 116
  holographic, 115–118
  longevity and, 118
  optical, 118–120
  volumetric, 115–116
Messenger RNA, 71, 142
Microbial engineering, 148–150
Micro-electromechanical systems (MEMs), 85, 114
Microfluidics, 157
Miller, Stanley, 49
MIT Artificial Intelligence Lab, 84–85, 148–150
Modulo arithmetic, 132
Moore, Tom & Ana, 121
Morphogenesis, 91
Morse, Daniel, 138
Motorola, 120
Munoz, Jose, 124
Musgrove, Ken, 23
Music, evolutionary, 22–23
Mutation operator, 16

Nagpal, Radhika, 93
Nanotechnology, 67–68
  biocomputation and, 77–79
  data storage and, 114–115
  manufacturing and, 69–71
  microbial engineering and, 148–150
  self-assembly and, 69–81
  Teramac and, 88–91
NASA, 120, 122, 133–134
National Science Foundation, 120
National Security Agency, 62
Natural Selection, Inc., 20
Navier-Stokes equations, 35
Neural networks
  applications of, 8–9

artificial intelligence and, 10–11
backpropagation algorithm and, 4, 8
binary response of, 5
biohardware and, 135–138
Church-Turing hypothesis and, 28
digital computers and, 9–10
digital Darwinism and, 14–15
games and, 20–22
genetic algorithms and, 20–22, 24, 141–149, 153
hidden layer of, 4, 6
intracellular signaling and, 139–148, 153–154
learning and, 2–3
mathematics and, 1–3
mechanisms of, 4–8
perceptrons and, 3
random noise and, 144
recurrent networks and, 3–4
synapses and, 2
synthetic cellular, 145–146
threshold value and, 2
weighted signals and, 5–8
Newton, Isaac, 13, 25
Nondeterministic polynomial time (NP), 60–61, 129

O excited state, 119
Ofria, Charles, 47
Optical memory, 118–120
"Organization of Behavior, The" (Hebbs), 2
Origin of the Species, The (Darwin), 13–14
Oscillators, 146–149

Parse tree, 17–18
Pathogens, 98–103
Pattern recognition, 8–9
Peptides, 105–106
Perceptrons, 3
Phages, 141–146
Photosynthesis, 120–122
  bacteriorhodopsin and, 114–115
  memory and, 115–116

Pitts, Walter, x, xii, 1, 3, 10, 27–28
Polymerase chain reaction (PCR), 58, 157
Population growth
  automata and, 32–36
  Tierra project and, 46–47
Prescott, David, 64
Programming. *See also* Algorithms
  amorphous computing and, 83–95
  a priori, 9
  DNA computation and, 51–65
  evolvable hardware and, 122–129
  genetic, 15–18
  Growing Point Language and, 91–95
Proteins, 139
  Cro, 145
  genetic switches and, 142–148
  memory and, 114–120
Proteomics, 139
Psychology, 2

Q excited state, 119
Quantum mechanics, ix, 25

Random noise, 144
Ray, Tom, 46–47
Rechenberg, Ingo, 15
Recombination, 64
Recurrent networks, 3–4
Reynolds, Craig, 45
Rivest, Ron, 53
RNA, 71, 142
Robotics, 11
  artificial life and, 42
  genetic algorithms and, 18–20
  GOLEM project and, 42, 44
  top-down approach to, 42
Rozenberg, Grzezorg, 64
RSA encryption, 53

Sakamoto, Kensaku, 62
Santa Fe Institute, 35, 39, 41, 135
Satisfiability problem (SAT), 61–62
Scanning tunneling microscope, 67
Schwefel, Hans-Paul, 15

Seeman, Ned, 79–80
Self-assembly
  adsorption and, 75
  biocomputation and, 77–79
  chemical synthesis and, 74–75
  definition of, 69
  DNA and, 71, 75–76, 79–80
  internal, 70–74
  noncovalent bonds and, 72–73
  SAMs and, 70, 75–76
  scale and, 67–71, 81
  thermodynamical equilibrium and, 72
Self-replication, 29–31
Self-tolerance
  computer virus detection and, 106–111
  human immune system and, 98–103
Shamir, Adi, 53
Shapiro, Lucy, 142, 144
Signals. *See* Neural networks
Sims, Karl, 42–43
Simulation Program with Integrated Circuit Emphasis (SPICE), 151–153
Single-instruction, multiple-data (SIMD) computation, 56
Sipper, Moshe, 122, 130, 135
Smart paint, 85–86
Soviet Union, 115
Spacial light modulator (SLM), 116
Speech production, 9
Splicing model, 53
Steingberg-Yfrach, Gali, 121
Stochastic effects, 144
Stoica, Adrian, 122, 129, 134
Submodules, 145–146
Sussman, Gerry, 84, 87, 93
Swiss Federal Institute of Technology (EPFL), 122, 130–132
Synapses, 2, 5
Synthetic biology, 149–150, 153

Taylor, Charles, 40
T cells, 102–103, 107
Teramac computer, 88–91

Thermodynamics, ix, 72
Thompson, Adrian, 127, 129
Thymine, 54–58
Tierra project, 46–47
Tilings, 80
Toffoli, Tommaso, 33–34
Toggle switches, 146–148
Traveling Salesman Problem, 53–54,
    58–61, 79
Turing, Alan, x, 27–28, 80
Tyrrell, Andy, 131, 133–135

Ulam, Stanislaw, 31
Ultraviolet light, 141–142, 144–145,
    147

van der Waals forces, 72
Viruses
  biohardware and, 133–135
  computer immune systems and, 97,
    106–111, 133–135
  lambda, 141–146
VLSI systems, 136–138
Volumetric memory, 115–116
von Neumann, John, 1, 40–41,
    156
  amorphous computing and, 83–84,
    93, 95
  automata and, 25–26, 28–31
  Church-Turing hypothesis and, 28
  fault tolerance and, 83–84
  Game of Life and, 32–33
  Hixon Symposium and, x–xii, 26

Wang, Hao, 80
Watson, James, xi, 31
Weighted signals, 5–8
Weinberg, Steven, 36
Weiss, Ron, 149, 153
Werbos, Paul, 8
Wet lab experimentation, 53
White cells, 99–100
Whitesides, George, 71
Winfree, Erik, 80
Wolfram, Stephen, 34
Woodbury, Neil, 120–121

World Economic Forum, 158

X-ray diffraction, 31, 86

Zhang, Shuguang, 76
Zygotes, 140–141